WiMAX

Applications

The
WiMAX
Handbook

WiMAX: Technologies, Performance Analysis, and QoS
ISBN 9781420045253

WiMAX: Standards and Security
ISBN 9781420045237

WiMAX: Applications
ISBN 9781420045474

The WiMAX Handbook
Three-Volume Set
ISBN 9781420045350

WiMAX

Applications

Edited by
**SYED AHSON
MOHAMMAD ILYAS**

CRC Press
Taylor & Francis Group
Boca Raton   London   New York

CRC Press is an imprint of the
Taylor & Francis Group, an informa business

CRC Press
Taylor & Francis Group
6000 Broken Sound Parkway NW, Suite 300
Boca Raton, FL 33487-2742

© 2008 by Taylor & Francis Group, LLC
CRC Press is an imprint of Taylor & Francis Group, an Informa business

No claim to original U.S. Government works
Printed in the United States of America on acid-free paper
10 9 8 7 6 5 4 3 2 1

International Standard Book Number-10: 1-4200-4547-4 (Hardcover)
International Standard Book Number-13: 978-1-4200-4547-5 (Hardcover)

Library of Congress Cataloging-in-Publication Data

Ahson, Syed.
    WiMAX : applications / Syed Ahson and Mohammad Ilyas.
        p. cm.
    Includes bibliographical references and index.
    ISBN 978-1-4200-4547-5 (alk. paper)
    1. Wireless communication systems. 2. Broadband communication systems.
    3. IEEE 802.16 (Standard) I. Ilyas, Mohammad, 1953- II. Title.

    TK5103.2.A4316 2008
    621.384--dc22   2007012502

Visit the Taylor & Francis Web site at
http://www.taylorandfrancis.com

and the CRC Press Web site at
http://www.crcpress.com

Contents

Preface

The demand for broadband services is growing exponentially. Traditional solutions that provide high-speed broadband access use wired access technologies, such as traditional cable, digital subscriber line, Ethernet, and fiber optic. It is extremely difficult and expensive for carriers to build and maintain wired networks, especially in rural and remote areas. Carriers are unwilling to install the necessary equipment in these areas because of little profit and potential. WiMAX will revolutionize broadband communications in the developed world and bridge the digital divide in developing countries. Affordable wireless broadband access for all is very important for a knowledge-based economy and society. WiMAX will provide affordable wireless broadband access for all, improving quality of life thereby leading to economic empowerment.

Broadband wireless access is as important as waterways, railroads, and interstate highways of an earlier era. Broadband wireless access technical solutions and products have been available for some time. These technologies have primarily focused on providing high data rate connectivity wirelessly between fixed stationary sites. These technical solutions are proprietary in nature and suffer from poor interoperability with other broadband wireless access products and high cost due to the lack of economy of scale. The IEEE 802.16 BWA technology family, referred to as worldwide interoperability for microwave access (WiMAX), intends to provide a standardized broadband wireless access solution. WiMAX has a strong base of standardization and industry support that provides a strong evolutionary path of its capabilities. The IEEE 802.16 specifications continue to evolve and expand in capabilities in support of the evolving vision of WiMAX usage and deployment.

WiMAX enables wireless broadband access anywhere, anytime, and on virtually any device and has generated unparalleled interest within the wireless networking community. WiMAX is the next step in the mobile technology evolution path; it competes with IEEE 802.11-based WLAN technology, broadband residential Internet technologies such as digital subscriber line and cable and third-generation cellular technologies. WiMAX offers numerous advantages, such as improved performance and robustness, end-to-end IP-based network, secure mobility, and broadband speeds for voice, data, and video.

The WiMAX handbook provides technical information about all aspects of WiMAX. The areas covered in the handbook range from basic concepts to research-grade material including future directions. The WiMAX handbook captures the current state of wireless local area networks, and serves as a source of comprehensive reference material on this subject. The WiMAX handbook consists of three volumes: *WiMAX: Applications*; *WiMAX: Standards and Security*; and *WiMAX: Technology, Performance Evaluation, and QoS*. It has a total of 32 chapters authored by experts from around the world. *WiMAX: Applications* includes 10 chapters authored by 21 experts from around the world.

Chapter 1 (WiMAX Past, Present, and Future: An Evolutionary Look at the History and Future of Standardized Broadband Wireless Access) describes the logical architecture of IEEE 802.16 and summarizes technology specifications. An example of an IEEE 802.16 network comprising core network, base station, and subscriber station is presented. Evolving usage scenarios for "fixed" and "mobile" WiMAX and role of competing technologies are examined in detail.

Chapter 2 (Overview of WiMAX Standards and Applications) introduces some of the main IEEE 802.16 family standards (802.16a, 802.16.2-2001, 802.16-2004, 802.16.2-2004, 802.16c, 802.16e, and 802.16f-2005). The application of WiMAX for rural area broadband services is analyzed. An example of last-mile access for providing high-speed access to buildings using WiMAX is given. A comprehensive list of WiMAX applications for video surveillance, automatic teller machines, online gaming, multimedia communication, medical applications, vehicular voice and data, sensor networks, telematics, and telemetry is included.

Chapter 3 (WiMAX Technology for Broadband Wireless Communication) compares WiMAX to Wi-Fi (IEEE 802.11a/b/g) with respect to coverage area, bandwidth, spectrum, technology, network deployment, and applications. Advantages of WiMAX technology such as high capacity, quality-of-service, flexible architecture, mobility, improved user connectivity, robust carrier class operation, scalability, nonline-of-sight connectivity, cost-effectiveness, and fixed and nomadic access are discussed. Cellular, military, medical, security, disaster recovery, public safety, and campus connectivity applications of WiMAX are described in detail.

Chapter 4 (VoIP over WiMAX) studies the feasibility of supporting VoIP over WiMAX and discusses a combination of techniques that can be adopted not only to enhance the performance of VoIP but also to support more numbers of VoIP calls. A simplified VoIP system architecture is presented. Performance of VoIP calls is studied with respect to an ITU-T E-Model *R*-score that combines different aspects of voice quality impairment.

Chapter 5 (WiMAX Technology for Home Access) presents issues related to the use of WiMAX technology as an alternative to provide broadband access to residential users. The feasibility of WiMAX for home access is analyzed from different perspectives, including data rate, frequency, coverage, and cell planning.

Chapter 6 (WiMAX Enables Cyber Extension to Rural Communities) illustrates the advantages of WiMAX as a last-mile solution and evaluates the potential of WiMAX as a 4G technology of the future. WiMAX with Wi-Fi as an application is presented as a solution to enable rural information and communication technology infrastructure. Economic and technical advantages of WiMAX are summarized. Analysis done in this chapter shows that WiMAX will win in the marketplace and will capitalize on this, and further supports the proliferation of VoIP devices and IP-based services.

Chapter 7 (WiMAX over GSM for Basic IP Access in African Rural Areas) shows that basic but affordable Internet protocol connectivity can be provided to rural communities by using spare capacity on GSM networks to carry WiMAX traffic. In general, rural areas in Africa are seen as unprofitable by operators and hence these areas do not benefit from typical wired Internet access. On the contrary, the global system for mobile communications has thoroughly penetrated Africa and in many cases unutilized capacity exists in rural areas. Since the main problem with wireless local area network in Africa is not the last mile, but rather finding a way to connect the wireless access point to an existing backbone network, a solution to integrate WiMAX with GSM is proposed.

Chapter 8 (Applications of Wi-Fi/WiMAX Technologies in the Emerging World) proposes a strategic wireless framework to address challenges in three different sectors of a developing country. Deployment and implementation of an affordable communications infrastructure with emerging wireless technologies are the first steps toward narrowing the digital divide. This chapter concludes that information and communication technology backed by modern wireless technologies will take any developing country into a new age of information economy and wealth creation.

Chapter 9 (Connectivity and Load Distribution in WiMAX-Based Multihop Backhaul Networks) examines the backhaul requirements of a large fixed wireless network providing high-speed data service to customer premises. Distribution of access point load is investigated and the required capacity of access-point-to-access-point and access-point-to-gateway links is characterized such that the occurrence of overload conditions is limited. Defining the service requirements of a single customer as a "unit load," the distribution of the load supported by a single access point is calculated.

Chapter 10 (Providing QoS to Real and Interactive Data Applications in WiMAX Mesh Networks) considers the problem of centralized routing and scheduling for IEEE 802.16 mesh networks so as to provide quality of service to real and interactive data applications. This chapter presents scheduling algorithms that provide per flow QoS guarantees while utilizing the network resources efficiently. Admission control policies, which ensure that sufficient resources are available, are discussed.

The targeted audience for the handbook includes professionals who are designers and planners for WiMAX networks, researchers (faculty members and graduate students), and those who would like to learn about this field.

The handbook is expected to have the following specific salient features:

- To serve as a single comprehensive source of information and as reference material on WiMAX networks
- To deal with an important and timely topic of emerging communication technology of today, tomorrow, and beyond

- To present accurate, up-to-date information on a broad range of topics related to WiMAX networks
- To present the material authored by the experts in the field
- To present the information in an organized and well-structured manner

Although the handbook is not precisely a textbook, it can certainly be used as a textbook for graduate and research-oriented courses that deal with WiMAX. Any comments from the readers will be highly appreciated.

Many people have contributed to this handbook in their unique ways. The first and the foremost group that deserves immense gratitude is the group of highly talented and skilled researchers who have contributed 32 chapters to this handbook. All of them have been extremely cooperative and professional. It has also been a pleasure to work with Nora Konopka, Helena Redshaw, Jessica Vakili, and Joette Lynch of Taylor & Francis and we are extremely gratified for their support and professionalism. Our families have extended their unconditional love and strong support throughout this project and they all deserve very special thanks.

Syed Ahson
Plantation, FL, USA

Mohammad Ilyas
Boca Raton, FL, USA

Editors

Syed Ahson is a senior staff software engineer with Motorola Inc. He has extensive experience with wireless data protocols (TCP/IP, UDP, HTTP, VoIP, SIP, H.323), wireless data applications (Internet browsing, multimedia messaging, wireless e-mail, firmware over-the-air update), and cellular telephony protocols (GSM, CDMA, 3G, UMTS, HSDPA). He has contributed significantly in leading roles toward the creation of several advanced and exciting cellular phones at Motorola. Prior to joining Motorola, he was a senior software design engineer with NetSpeak Corporation (now part of Net2Phone), a pioneer in VoIP telephony software.

Syed is a coeditor of the *Handbook of Wireless Local Area Networks*: *Applications, Technology, Security, and Standards* (CRC Press, 2005). Syed has authored "Smartphones" (International Engineering Consortium, April 2006), a research report that reflects on smartphone markets and technologies. He has published several research articles in peer-reviewed journals and teaches computer engineering courses as adjunct faculty at Florida Atlantic University, Florida, where he introduced a course on smartphone technology and applications. Syed received his BSc in electrical engineering from India in 1995 and MS in computer engineering in July 1998 at Florida Atlantic University, Florida.

Dr. Mohammad Ilyas received his BSc in electrical engineering from the University of Engineering and Technology, Lahore, Pakistan, in 1976. From March 1977 to September 1978, he worked for the Water and Power Development Authority, Pakistan. In 1978, he was awarded a scholarship for his graduate studies and he completed his MS in electrical and electronic engineering in June 1980 at Shiraz University, Shiraz, Iran. In September 1980, he joined the doctoral program at Queen's University in Kingston, Ontario, Canada. He completed his PhD in 1983. His doctoral research was about switching and flow control techniques in computer communication networks. Since September 1983, he has been with the College of Engineering and Computer Science at Florida Atlantic University, Boca Raton, Florida, where he is currently associate dean for research and industry relations. From 1994 to 2000, he was chair of the Department of Computer Science and Engineering. From July 2004 to September 2005, he served as interim associate vice president for research and graduate studies. During the 1993–1994 academic year, he was on his sabbatical leave with the Department of Computer Engineering, King Saud University, Riyadh, Saudi Arabia.

Dr. Ilyas has conducted successful research in various areas including traffic management and congestion control in broadband/high-speed communication networks, traffic characterization, wireless communication networks, performance modeling, and simulation. He has published one book, eight handbooks, and over 150 research articles. He has supervised 11 PhD dissertations and more than 37 MS theses to completion. He has been a consultant to several national and international organizations. Dr. Ilyas is an active participant in several IEEE technical committees and activities.

Dr. Ilyas is a senior member of IEEE and a member of ASEE.

Contributors

Dan Avidor
Bell Laboratories, Lucent
   Technologies
Alcatel-Lucent
Holmdel, New Jersey

Jack L. Burbank
Applied Physics Laboratory
The Johns Hopkins University
Laurel, Maryland

Albert Butare
Ministry of Infrastructure
Kachiyuru, Kigali, Rwanda,
   Central Africa

Damien Chatelain
University of Technology
Pretoria, South Africa

Mainak Chatterjee
University of Central Florida
Orlando, Florida

Giselle M. Galván-Tejada
CINVESTAV-IPN
Mexico City, Mexico

Vinoth Gunasekaran
Stevens Institute of Technology
Hoboken, New Jersey

Neena Gupta
Punjab Engineering College
Deemed University
Chandigarh, India

Fotios C. Harmantzis
Stevens Institute of Technology
Hoboken, New Jersey

William T. Kasch
Applied Physics Laboratory
The Johns Hopkins University
Laurel, Maryland

Gurjit Kaur
University Institute of Engineering
   and Technology
Chandigarh, India

G. Senthil Kumaran
Kigali Institute of Science and
   Technology
Kigali, Rwanda, Central Africa

Sayandev Mukherjee
Marvell Semiconductor
Santa Clara, California

Kumbesan Sandrasegaran
University of Technology
Sydney, Australia

K.R. Santhi
Kigali Institute of Science and
   Technology
Kigali, Rwanda, Central Africa

Shamik Sengupta
University of Central Florida
Orlando, Florida

Vinod Sharma
Indian Institute of Science
Bangalore, India

Harish Shetiya
Ittiam Systems
Bangalore, India

Erickson Trejo-Reyes
Nextel de Mexico
Mexico City, Mexico

Barend J. van Wyk
University of Technology
Pretoria, South Africa

Leijia Wu
University of Technology
Sydney, Australia

1

WiMAX Past, Present, and Future: An Evolutionary Look at the History and Future of Standardized Broadband Wireless Access

Jack L. Burbank and William T. Kasch

CONTENTS

1.1   Introduction

Broadband wireless access (BWA) technical solutions and products have been available for some time. Historically, these technologies have been primarily focused on providing high data rate connectivity wirelessly between fixed stationary sites. Examples of these types of applications include building-to-building bridging and providing high-rate connectivity to remote sites, such as broadcast towers, where the installation of wired infrastructure is not viable. However, these technical solutions have historically been proprietary in nature and have suffered from several of the negative characteristics often accompanying proprietary solutions, including poor interoperability with other BWA products and high cost due to the lack of economy of scale.

The IEEE 802.16 BWA technology family, often referred to as world-wide interoperability for microwave access (WiMAX) or WirelessMAN, is intended to provide a standardized BWA solution to provide "broadband wireless to the masses" and is so anticipated that it has even been characterized by some as a threat to the long-term viability of several existing wireless technologies (including IEEE 802.11-based wireless local area network [WLAN] technology, broadband residential Internet technologies such as digital subscriber line [DSL] and cable), even viewed by some as a competitor to third-generation (3G) cellular technologies. Others view 802.16 as a powerful complementary technology to these various tools. Regardless, 802.16 is seen by many in the commercial wireless industry as a key enabling technology in the large-scale realization of the wireless Internet, providing a tool that may potentially allow service providers to deliver high data rates (i.e., tens of Mbps) to a variety of devices, such as handheld devices, and enabling an entire new generation of applications (e.g., handheld, high-resolution videophones).

The future success of WiMAX in the commercial marketplace and its potential emergence as a disruptive technology is still unknown. While WiMAX has certainly generated a high degree of excitement within the commercial wireless industry, the marketplace always proves to be the final judge—that which separates hype from reality. However, this determination is a complex matter that is a function of numerous factors. The goal of this chapter is to provide an overview of the evolution of WiMAX from three key perspectives: (1) usage case, (2) technology, and (3) standardization.

1.2   WiMAX—An Overview

WiMAX is based upon the IEEE 802.16 WMAN technology family, which provides specifications of the media access control (MAC) layer and the physical (PHY) layer. The 802.16 specification further subdivides the MAC sublayer into three sublayers: the convergence sublayer (CS), the common part sublayer (CPS), and the security sublayer. The CS aims to enable 802.16 to better accommodate the higher layer protocols placed above the MAC layer. The 802.16 specification assumes there will be two predominant types of traffic transported across the 802.16 network: ATM and IEEE 802.3 (Ethernet). Thus, there are two CS specifications: ATM and packet. The CS receives data frames from a higher layer and classifies the frame. On the basis of this classification, the CS can perform additional processing, such as payload header compression, before passing the frame to the MAC CPS. The CS also accepts data frames from the MAC CPS. If the peer CS has performed any type of processing, the receiving CS will restore the data frame before passing it to a higher layer. The CS is separate from the remainder of the 802.16 MAC such that vendors who wish to support other protocols can develop specialized CSs.

The CPS is the central piece of the 802.16 MAC, defining the medium access method (Figure 1.1). The CPS provides functions related to duplexing, network entry and initialization, framing, quality of service (QoS), and channel access. The security sublayer, also referred to as the privacy sublayer, has

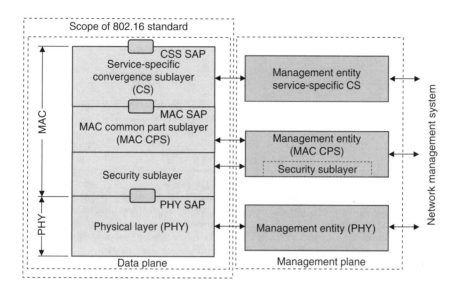

FIGURE 1.1
The logical architecture of IEEE 802.16.

been designed to meet two primary goals: providing subscribers with privacy across the wireless network and providing operators with strong protection from theft of service. The PHY layer then converts MAC layer frames into signals to be transmitted across the air interface. Consequently, the security sublayer has two component protocols: an encapsulation protocol and a privacy key management protocol.

The 802.16 technology family is actually composed of several distinct technology specifications. These specifications are summarized in Table 1.1.

The term "WiMAX" is a marketing term that has become synonymous with 802.16-based BWA networks in much the same manner as "wireless fidelity" or "Wi-Fi" has become synonymous with IEEE 802.11-based WLANs. The WiMAX Forum was formed in April 2001 as a nonprofit international organization to certify conformance and interoperability of products on the basis of the IEEE 802.16 and ETSI HIPERMAN standards. This forum is also heavily involved as an advocate for 802.16 technology. It has now grown to include over 420 member companies.

The WiMAX-certified logo of the WiMAX Forum will be placed on the package of certified products, and is envisioned to become a key criterion for market viability in the same way that the Wi-Fi-certified logo of the Wi-Fi Alliance is a key criterion for market viability of 802.11 WLAN products.

1.2.1 The WiMAX Standard

There is typically much confusion regarding the "WiMAX standard." WiMAX is not a standard. WiMAX is a marketing term trademarked by the WiMAX

TABLE 1.1

Summary of Various 802.16 Technology Specifications

Specification	Reference	Year of Ratification	Description
802.16	1	2001	MAC and PHY definition for fixed broadband wireless access in the 10–66 GHz frequency bands.
802.16a	2	2003	Amendment to the original specification. Contains new PHY definitions for the 2–11 GHz frequency bands. Also includes mesh network modes of operation.
802.16c	3	2002	System profiles for 10–66 GHz operations.
802.16d	4	2004	Contains 802.16, 802.16a, and various MAC enhancements. Commonly referred to as 802.16-2004. Considered the base 802.16 fixed broadband wireless specification.
802.16e	5	2006	Amendment to the 802.16d specification to provide explicit support for mobility. Incorporates WiBRO. Commonly referred to as 802.16-2005. Considered the base 802.16 mobile broadband wireless specification.
802.16f	6	2005	802.16 management information base.
802.16g	N/A	In progress	Network management (management plane control procedures).
802.16h	N/A	In progress	Coexistence in license-exempt frequency bands.
802.16i	N/A	In progress	Mobile management information base.
802.16j	N/A	In progress	Multihop relay specification.
802.16k	N/A	In progress	802.16 MAC-layer bridging.
802.20	N/A	In progress	Mobile broadband wireless access standards group. Initially formed as a standards group within the 802.16 Working Group, it consisted of a group of individuals who wished to develop a new technology focused solely on mobility. No other relation to WiMAX, other than perhaps competitive.
WiBRO			Korean wireless broadband standard incorporated into the 802.16e (802.16-2005) standard.

Forum to describe 802.16-based technology. 802.16 is a technology standard. However, it is not uncommon for 802.16 and WiMAX to be referred to as separate standards. The WiMAX standard refers to the set of capabilities within 802.16 that the WiMAX Forum will test against when performing conformance and interoperability testing in its equipment certification process. In this sense, the WiMAX Forum will indeed have significant impact on what functionality within the 802.16 standard will be brought to market by vendors, but it in itself is not a standard. Rather, the WiMAX standard refers to the subset of 802.16 capabilities that are likely to experience widespread implementation.

1.2.2 Current WiMAX Product Market

It is important to note at this point that there are relatively few WiMAX-certified products currently available on the market. Rather, many so-called WiMAX products currently available are proprietary in nature. The WiMAX Forum opened its laboratory only in mid-2005 to begin certifying conformance and interoperability for fixed equipments at 3.5 and 5.8 GHz, with additional spectrum channels to follow; the first certifications were issued only in early 2006 for fixed WiMAX products and no mobile WiMAX products are yet certified. However, many vendors already offer products that they advertise as WiMAX-ready (both fixed and mobile). These products are based on potentially proprietary technologies but are advertised as capable of being brought into WiMAX conformance via software upgrade only. Whether an organization or individual decides to deploy this equipment with the future goal of interoperability with WiMAX-certified equipments, is an individual choice, and is largely a function of trust and confidence in the equipment vendor.

1.2.3 A Brief Overview of IEEE 802.16 Networking

The 802.16 network architecture is predicated on the presence of fixed infra-structural sites. In fact, the architectural model of 802.16 is similar to the model employed within cellular telephone networks. Each 802.16 coverage area consists of one base station (BS) and one or more subscriber stations (SSs). BSs provide connectivity to core networks (CNs), whereas the SS is the suite of the equipment at the customer location, or customer premises equipment, which provides access for the end user into the broadband wireless network. A single 802.16 coverage area is depicted in Figure 1.2. The architecture depicted in this figure represents a single cell of network coverage. These 802.16 cells

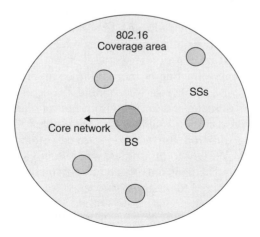

FIGURE 1.2
The 802.16 coverage area.

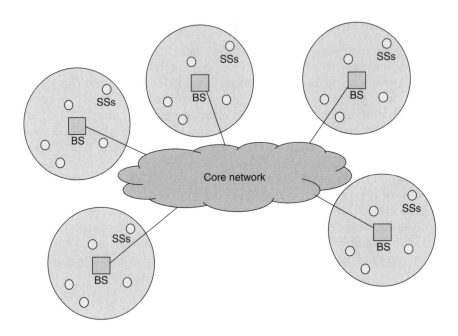

FIGURE 1.3
An example 802.16 network.

can then be grouped together to form a larger 802.16 network, where the BS sites are interconnected via a CN, as depicted in Figure 1.3. In the 802.16 model, channel access is highly centralized; the BS is in complete control over how and when SSs access the wireless medium.

Transmissions may be point-to-point (PTP), point-to-multipoint (PMP), or point-to-consecutive-point (PTCP) in nature, where PTCP involves the creation of a closed loop through multiple PTP connections. In addition, 802.16a provides for a mesh networking capability in which SSs can act as routers, relaying data to nodes that may not have line-of-sight (LOS) connectivity with the BS. BSs typically employ one or more wide-beam antennas that may be partitioned into several smaller sectors, where all sectors sum to complete 360° coverage. This is analogous to BSs within the cellular model. SSs typically employ highly directional antennas that are pointed toward the BS. This is a significant departure from the model employed within cellular communications or the 802.11 WLAN communities, where low-gain, omnidirectional antennas are employed. This is one of the key reasons 802.16 achieves such higher data rates compared to other technologies.

The BS-to-SS link is referred to as the downlink. The SS-to-BS link is referred to as the *uplink*. The proper routing of traffic to a BS is a function of the CN, which is not explicitly defined within the 802.16 specification. In fact, the 802.16 specification has provisions to accommodate a multitude of existing or future CN technologies. This CN is analogous to the DS of 802.11 networks.

1.3 WiMAX—Evolving Usage Cases

Any successful technology must fit a key need in the marketplace. That is, a technology must have a compelling usage case, a "killer niche" that it can fit better than alternatives. And here, niche does not imply any type of limited size of scale, but rather is viewed as analogous to the "killer app." This is often why the position of "first-to-market" is so attractive, because it eliminates competition from viable alternatives, leaving that technology as the only choice to meet the compelling usage case. One needs to look no further than the wildly successful IEEE 802.11 WLAN technology family to see the importance of a killer niche. IEEE 802.11 found a niche in the marketplace for localized hot-spot wireless connectivity, the ability to replace network cables within localized regions, and has filled that niche with a family of highly capable technologies that now approaches ubiquity. This original usage case of IEEE 802.11 WLANs was simple and narrow in focus. Only now do later revisions of the technology address more sophisticated usage cases, such as QoS, mesh networking, and roaming.

Clearly, it is quite important for WiMAX to have a killer niche that it can satisfy to enjoy long-term success in the marketplace. However, an agreement on exactly what that compelling usage case is or should be has been difficult to come to, even within the WiMAX community. Even up to this point in time, the vision for how WiMAX will be or should be employed is arguable.

Certainly, the dominant usage case scenarios have evolved over time as WiMAX has continued to evolve from both a marketing and a technology perspective. The original 802.16 specification [1] was clearly oriented toward providing high-rate, PTP, LOS connectivity between fixed platforms. Here, the driving usage scenario was that of interconnecting locations that do not lend themselves to cabled solutions. A classic example of this usage scenario is that of the remotely located transmission tower that is wirelessly backhauled to a fixed location attached to a larger wired network. There was, and still is, a legitimate market within this problem space. However, this market was continually hampered by poor interoperability between proprietary solutions. The goal of creating an interoperable technology to fit this niche was the original inspiration of the 802.16 specification, the envisioned backhaul technology of this problem space. In this envisioned usage case, the primary competition to WiMAX is proprietary solutions. It is clear that, in the long term, a standardized technology with strong industry support, such as WiMAX, would enjoy tremendous success. However, this is the narrowest of envisioned problem spaces for which WiMAX is often considered a candidate.

After the finalization of the original 802.16 technology standard, the scope of envisioned WiMAX usage scenarios was significantly expanded. Originally, WiMAX was viewed as a PTP, LOS backhaul technology, envisioned to provide wireless bridging between fixed locations within network infrastructure.

The primary expansion of scope was to that of the direct support of end-user networks. Here, WiMAX was envisioned to serve a role within the internet service providers (ISP) problem space, interconnecting end-user networks (e.g., homes) with network infrastructure. This expansion of scope makes very good logical sense, particularly given the still fixed nature of network nodes. Here, the creators of WiMAX had created a wireless technology capable of delivering very high data rates over very long distances. Indeed, this would seem an ideal technology to apply to the problem space of the residential wireless local loop, where low-rate wired infrastructure often limits the types of capabilities that can be enjoyed by the residential consumer. The resultant technology brought to bear in this problem space was the IEEE 802.16a specification [2], leading to the eventual IEEE 802.16d specification, which unified the original 802.16 and 802.16a specifications. This 802.16d specification is also often referred to as the IEEE 802.16-2004 specification and is the basis for fixed WiMAX. However, this problem space contains much stiffer competition than the original scope of employment in the form of both wire-line and wireless technologies. Technologies such as DSL and cable modem are already firmly entrenched within this market space, already enjoying significant consumer bases. Additionally, there are other wireless technologies also contending for this market, such as CDMA2000. The marketplace will ultimately determine the level of success any of these technologies will experience. However, current conventional wisdom is that technologies such as WiMAX are more likely to enjoy success in this problem space within markets that are not yet fully developed (e.g., third-world markets) or in regions where other forms of ISPs do not have a strong presence (e.g., rural areas within developed countries).

The final evolution of the WiMAX usage scenario came in the form of mobility support. Here, the envisioned scenario has WiMAX serving as the air interface for the actual radio access network, where both fixed and mobile users access the WiMAX network. The developers of the technology had created a technology capable of reasonably high data rates at reasonably long ranges. If this technology could now be augmented to support the case of the mobile users, then WiMAX could serve as a viable candidate for wide-area connectivity. This usage case is the driving scenario behind the creation of the IEEE 802.16e technology standard [5], also referred to as IEEE 802.16-2005, the basis of mobile WiMAX. This market arguably presents the stiffest competition of all envisioned usage scenarios. Within this space, there are two potential deployment scenarios: (1) employment of the WiMAX air interface by incumbent wireless service providers (WSPs) and (2) employment of the WiMAX air interface by new-entry WSPs. Incumbent cellular providers have invested enormous amounts of capital expenditures to reach current level of capabilities that will not be easily equaled or surpassed by any new-entry technology. Even next-generation cellular technologies, such as 3GPP and 3GPP2, have experienced relatively slow deployment as cellular service providers have not been quick to embrace these technologies over older technologies such as GSM. It is unclear whether existing incumbents would embrace a technology such as WiMAX. It is reasonable to expect that WiMAX would be a compelling technology for a new-entry WSP. However, a new-entry WSP faces enormous challenges in the marketplace. Spectrum is expensive, infrastructure is expensive, and reaching the economy of scale required to drive down equipment and service costs to a competitive level is very challenging. Originally, one of the key enablers for this type of usage scenario was that of offering wireless service in unlicensed frequency bands. Thus was born the promise of "anybody can be a service provider" and brought great hope that a new-entry WSP could compete with incumbents. However, this is not realistic given that the WiMAX Forum has no certification profiles for unlicensed (5.8 GHz) mobile WiMAX. Thus, the majority of deployments, particularly those within the United States, will utilize licensed spectrum. Indeed, it is envisioned that mobile WiMAX will make its largest impact within this problem space in the United States within the 2.5 GHz multichannel multipoint distribution service (MMDS) frequency bands by incumbent WSPs.

It can be seen that the evolution of WiMAX usage cases can be characterized by an increasing scope and scale, along with much stiffer competition from other technologies. This corresponds to the increasing confidence that the proponents of WiMAX have in the technology they have developed. Certainly, there is a lot of excitement regarding WiMAX, and there is a lot of momentum building that favors rapid WiMAX adoption and deployment. This is evidenced by the escalating grandeur of the envisioned usage scenarios by the community. However, admittedly arguable and speculative, WiMAX will face significant difficulties in emerging as a serious competitor to 3G technologies. This is due to several complicating factors: (1) the evolution of other "competitive" technologies and (2) the lack of a "killer app" in the mobile data networking space. The evolution of 3GPP to high-speed downlink packed access renders the increased data rates of WiMAX merely an incremental increase. It is unclear if this incremental increase in data rate will motivate existing service providers to migrate to WiMAX. It should also be noted that WiMAX is only an air interface replacement, and that there remains the issue of deploying and maintaining a CN. Furthermore, it is unclear if WiMAX will mount a significant challenge to 802.11-based WLANs or residential broadband technologies such as DSL and cable. 802.11 is evolving quickly to several hundred Mbps solutions, and has a rapidly evolving suite of technologies for aspects such as mobility and roaming support. Both 802.11 and 3G have a several-year lead time to market over WiMAX. Residential broadband technologies such as DSL and cable are firmly entrenched in the market. For these reasons, it is envisioned by the authors that WiMAX will likely remain a complementary technology to these technologies, or shall remain a niche technology serving very specific usage cases. Most notably are (1) the original usage case of backhaul connectivity, (2) wireless local loop service to fixed locations in underdeveloped regions, and (3) mobile radio access in developing regions.

Another issue facing mobile WiMAX is that of the lack of a "killer app" that draws the masses to fully mobile data networking. Certainly, this is not only an issue unique to WiMAX but also an important issue facing cellular service providers. Despite the significant emergence of wireless networking technologies, it is still unclear whether there is an overwhelming market for truly mobile data networking. There is certainly a strong marketplace for nomadic mobility, the ability to move from one location to another with connectivity achievable from either location. However, it is not yet clear whether there is an overwhelming demand for data networking that can provide seamless connectivity while on the move. There are certainly examples where the motivation is quite strong. Military networks certainly need to be capable of operating in a seamless fashion while on the move. There are also the classical mobile data networking scenarios of public transit vehicles providing network services (e.g., train). However, these usage cases do not necessarily constitute a mainstream need for on-the-move network connectivity. Rather, despite being quite arguable, nomadic mobility is likely still the driving demand from consumers. This is also often referred to as portability. It should be noted that fixed WiMAX has already experienced significant deployments in which nomadic mobility has been demonstrated. However, as always, the marketplace will make the final determination as to which usage scenarios are viable, and which are not. All else is speculation. One development to watch closely, which could provide significant insight into the viability of WiMAX in the mobile radio access network problem space, is the ongoing deployment of WiBRO in South Korea in the 2.3 GHz band.

1.4 WiMAX—Evolution of the Technology

As the envisioned usage scenario has evolved over time, so has evolved the technological basis of WiMAX. The IEEE 802.16 technical specification has now evolved through three generations:

- IEEE 802.16: High data rate, high-power, PTP, LOS, fixed SSs
- IEEE 802.16-2004: Medium data rate, PTP, PMP, fixed SSs
- IEEE 802.16-2005: Low-medium data rate, PTP, PMP, fixed or mobile SSs

The first generation of IEEE 802.16 operates in microwave frequencies (hence the name) 10–66 GHz and utilizes single-channel (SC) modulation as it assumes LOS propagation is required for communications. This WirelessMAN-SC physical layer can employ QPSK, 16-QAM, or 64-WAM modulation, adaptively changing on the basis of channel conditions. The original 802.16 specification operates with channel bandwidths of 20–25 MHz in the United States and 28-MHz channel bandwidths in Europe.

This technology employs highly directional antennas and high-power levels within licensed frequency bands to achieve simultaneously high data rates and long ranges. Security mechanisms within the original specification are somewhat rudimentary, with a reliance on antenna directionality to mitigate intrusion. As can be seen, this technology is well suited to a fixed point-to-fixed point backhaul type of application.

The IEEE 802.16-2004 specification amends the original specification to operate in the 2–11 GHz, both licensed and license-exempt. This frequency band of operations, which was first addressed in the IEEE 802.16a specification, assumes non-LOS communications. This specification provides a total of three air interfaces:

    a. WirelessMAN-SC2—single-carrier modulation.

    b. WirelessMAN-OFDM—OFDM modulation with a 256-point fast fourier transform (FFT) with TDMA channel access.

    c. WirelessMAN-OFDMA—OFDM is employed with a 2048-point FFT. Multiple access is provided by addressing a subset of carriers to individual receivers.

In addition to the forward error control (FEC) coding employed in the original specification, the 2–11 GHz PHY specification also allows for the use of automatic retransmission requests as an optional capability. This technology incorporates numerous MAC-layer enhancements to the 802.16-2004 specification, including the support of multihop mesh networking to enable relaying between nodes to extend coverage areas of WiMAX BSs. This technology often operates using sectored omnidirectional antennas, decreasing dependence on precise antenna pointing and increasing the ability to provide entire coverage areas of service. Furthermore, operation in the 2–11 GHz frequency band allows for adaptive antenna beam-forming techniques to improve interference and scalability performance. Numerous security enhancements, such as two-way authentication, were included in this update to the original specification. It is readily apparent that this technology was certainly designed for the wireless local loop type of application.

The IEEE 802.16-2005 specification was developed with one primary goal: the support for a large number of mobile users. A key enhancement of the IEEE 802.16-2005 specification is the employment of scalable OFDMA (as opposed to the nonscalable version employed in the fixed WiMAX specification), which technology proponents argue makes the technology highly robust to network congestion and highly graceful degradation in the presence of interference. Other key enhancements include the introduction of several state-of-the-art technologies, such as hybrid automatic retransmission request, advanced FEC coding schemes such as turbo codes and low-density parity check codes, and multiple-input multiple-output.

In general, the technological evolution of WiMAX has traded capacity and range for mobility support and scalability. Figure 1.4 illustrates the basic

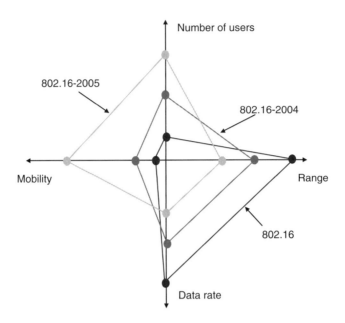

FIGURE 1.4
Evolution of WiMAX technologies.

capabilities of these various forms of WiMAX. From this figure, the trend is clearly toward compromising range and data rate for scalability and mobility support.

An important point to make here is that these various flavors are not compatible with one another. That is, an 802.16-2004 BS cannot interoperate with an 802.16-2005 SS, and vice versa. This could significantly constrain 802.16 deployment in the future. However, major chip manufacturers have already announced dual-mode chipsets that will support both standards. Thus, there will likely emerge products that can interoperate with both 802.16-2004 and 802.16-2005 networks. Unfortunately, there remains numerous regulatory and coexistence issues that complicate if not prohibit heterogeneous Fixed and Mobile WiMAX networks.

1.5   Relevant Standardization Activities

Another key for any successful technology is a strong evolution path. Certainly, this has become a key attractive feature of IEEE 802.11 WLAN technology. The IEEE 802.11 working group is actively working to address numerous issues and deficiencies in existing WLAN technologies. Indeed, who wants to invest enormous amounts of capital resources on a network infrastructure that is going to become obsolete quicker than necessary? Rather, one wishes to acquire a solution that will grow and evolve with the needs of the user. Thus, it is important to consider the strength of the industrial support of a technology and the amount of standardization activity to ensure competition among vendors.

This is an area in which WiMAX is very strong. WiMAX indeed enjoys enormous industry backing. The WiMAX Forum is a consortium of hundreds of companies, all proponents of the technology. Major industry players such as Intel (which certainly played a prominent role in the success of IEEE 802.11 with the success of its embedded Centrino chipset) are active in this forum and in the development of WiMAX devices and technology standards. The WiMAX Forum currently operates eight working groups: application, certification, global roaming, marketing, networking, regulatory, service provider, and technical. Each of these working groups are chartered to address particular aspects of the WiMAX technology to help ensure its successful adoption and deployment. For example, the networking working group creates networking specifications beyond that defined within the 802.16 specification as necessary to support fixed, nomadic, portable, and mobile WiMAX systems.

As can be seen from Table 1.1, the IEEE 802.16 working group continues to be quite active in the development of refinements to the IEEE 802.16 technology base of WiMAX. There are currently six active groups within the IEEE 802.16 working group, each working on a unique aspect of 802.16, such as specifications for 802.16 multihop relaying.

1.6 Conclusion

There are numerous factors that contribute to the success (or lack thereof) of a technology. WiMAX has generated a tremendous, almost unparalleled, amount of interest within the wireless networking community. Prior to its deployment, it was already being referred to as a disruptive technology. Time indeed will tell how disruptive it will become; it certainly has the potential to be landscape altering. WiMAX has a strong base of standardization and industry support that provides a strong evolutionary path of capabilities. Its technology base, the IEEE 802.16 specifications, has continued to evolve and expand in capabilities in support of the evolving vision of WiMAX usage and deployment. However, WiMAX faces very stiff competition from technologies such as 3GPP and 3GPP2, as well as expanding metropolitan-scale deployments of 802.11 WLANs. It should be very interesting to watch how the role of WiMAX now evolves within the emerging wireless Internet.

References

1. IEEE 802.16-2001, IEEE standard for local and metropolitan area networks—Part 16: Air interface for fixed broadband wireless access systems, 6 December 2001.

2. IEEE 802.16a-2001, IEEE standard for local and metropolitan area networks—Part 16: Air interface for fixed broadband wireless access systems—Amendment 2: Medium access control modifications and additional physical layer specifications for 2–11 GHz, 1 April 2003.
3. IEEE 802.16c-2001, IEEE standard for local and metropolitan area networks—Amendment 1: Detailed system profiles for 10–66 GHz, 15 January 2003.
4. IEEE 802.16-2004, IEEE standard for local and metropolitan area networks: Air interface for fixed broadband wireless access systems, 1 October 2004.
5. IEEE 802.16E-2005, IEEE standard for local and metropolitan area networks—Part 16: Air interface for fixed and mobile broadband wireless access systems amendment for physical and medium access control layers for combined fixed and mobile operation in licensed bands, 28 February 2006.
6. IEEE 802.16f-2005, IEEE standard for local and metropolitan area networks—Part 16: Air interface for fixed broadband wireless access systems—Amendment 1—Management information base, 1 December 2005.

2

Overview of WiMAX Standards and Applications

Leijia Wu and Kumbesan Sandrasegaran

CONTENTS

2.1   Overview of WiMAX

The demand for broadband services is growing sharply today. The traditional solutions to provide high-speed broadband access is to use wired access technologies, such as cable modem, digital subscriber line (DSL), Ethernet, and fiber optic. However, it is too difficult and expensive for carriers to build and maintain wired networks, especially in rural and remote areas. Broadband wireless access (BWA) technology is a flexible, efficient, and cost-effective solution to overcome the problems. The global deregulation of radio spectrum also encourages the development of BWA technologies. WiMAX is one of the most popular BWA technologies today, which aims to provide high-speed broadband wireless access for wireless metropolitan area networks (WMANs). The air interface standard, IEEE 802.16, commonly referred to as WiMAX, is a specification for broadband wireless communication standards developed for WMANs, which supports fixed, nomadic, portable, and mobile broadband accesses and enables interoperability and coexistence of BWA systems from different manufacturers in a cost-effective way. Compared to the complicated wired network, a WiMAX system only consists of two parts: the WiMAX base station (BS) and WiMAX subscriber station (SS), also referred to as customer premise equipments. Therefore, it can be built quickly at a low cost. Ultimately, WiMAX is also considered as the next step in the mobile technology evolution path. The potential combination of WiMAX and CDMA standards is referred to as 4G. This chapter gives an overview of the WiMAX standards and applications.

2.2   WiMAX Standards

The purpose of developing 802.16 standards is to help the industry to provide compatible and interoperable solutions across multiple broadband segments and to facilitate the commercialization of WiMAX products. Currently, WiMAX has two main variations: one is for fixed wireless applications (covered by IEEE 802.16-2004 standard) and another is for mobile wireless services (covered by IEEE 802.16e standard). Both of them are evolved from IEEE 802.16 and IEEE 802.16a, the earlier versions of WMAN standards. The 802.16 standards only specify the physical (PHY) layer and the media access control (MAC) layer of the air interface while the upper layers are not considered.

In the following sections, we will introduce some of the main IEEE 802.16 family standards.

2.2.1   802.16

The IEEE 802.16 standard (also known as the air interface for fixed broadband wireless access (FBWA) systems or IEEE WMAN air interface) is the first version of 802.16 family standards (published in April 2002). It specifies fixed broadband wireless systems operating in the 10–66 GHz licensed spectrum, which is expensive but there is less interference at the high-frequency band and more bandwidth is available. Because radio waves in this band are too short to penetrate buildings, the 802.16 standard is only used for line-of-sight (LOS) connections. Compared to nonline-of-sight (NLOS) connections, LOS links are not so flexible but are stronger and more stable against transmission errors. IEEE 802.16 is interoperable with other wireless networks, such as cellular systems and wireless local area networks (WLANs). In the following sections, the main features of 802.16 will be introduced.

2.2.1.1   Network Topology

802.16 defines two WiMAX network topologies: point-to-point (PTP) and point-to-multipoint (PMP). The PTP link refers to a dedicated link that connects only two nodes: BS and subscriber terminal. It utilizes resources in an inefficient way and substantially causes high operation costs. It is usually only used to serve high-value customers who need extremely high bandwidth, such as business high-rises, video postproduction houses, or scientific research organizations. In these cases, a single connection contains all the available bandwidth to generate high throughput. A highly directional and high-gain antenna is also necessary to minimize interference and maximize security.

Although PTP can be applied in the above special cases, it is too expensive for common customers. The PMP topology, where a group of subscriber terminals are connected to a BS separately (shown in Figure 2.1), is a better choice for users who do not need to use the entire bandwidth. Under PMP topology, sectoral antennas with highly directional parabolic dishes (each dish refers to a sector) are used for frequency reuse. The available bandwidth now is shared between a group of users, and the cost for each subscriber is reduced.

2.2.1.2   802.16 Protocol Stack

The 802.16 standard covers the lowest two layers in the OSI model: MAC layer and PHY layer (shown in Figure 2.2). The MAC layer is responsible for determining which SS can access the network and is further divided into three sublayers: service-specific convergence sublayer (CS), MAC common part sublayer (CPS), and security sublayer. The CS transforms incoming data received from the CS service access point (SAP) into MAC data packets. The transformation maps external network information into IEEE 802.16 MAC

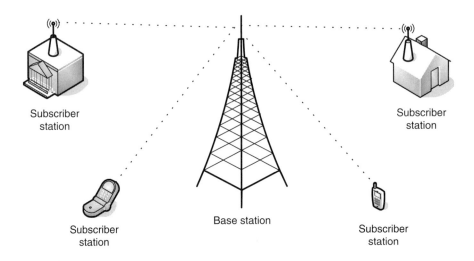

FIGURE 2.1
Point-to-multipoint WiMAX network topology.

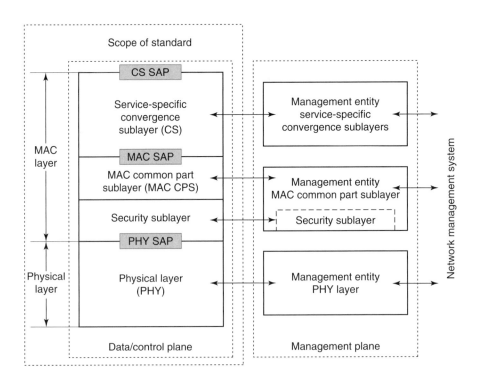

FIGURE 2.2
IEEE 802.16 protocol stack. (Reprinted with permission from IEEE Standard for Local and Metropolitan Area Networks Part 16: Air Interface for Fixed Broadband Wireless Access Systems, © IEEE 2002.)

information, such as service flow and connection identifier (CID). The current standard details two CS specifications: ATM CS and Packet CS. The CS is also responsible for preserving/enabling QoS and allowing bandwidth allocation.

The CPS is responsible for access control functionality, bandwidth allocation, connection establishment, and maintenance. Data, PHY control, and other management information are exchanged between the MAC CPS and PHY via the PHY SAP. The security sublayer is responsible for authentication, key exchange, and encryption.

IEEE 802.16 PHY is responsible for data transmission and reception. It is specified for the 10–66 GHz spectrum assuming LOS between the BS and the SS. IEEE 802.16 PHY supports wide channel bandwidth of 20, 25, or 28 MHz.

2.2.1.3 Modulation Technologies

IEEE 802.16 uses single-carrier modulation schemes in which all packets are sequentially transmitted through a single frequency carrier. Three modulation schemes are supported: QPSK (quadrature phase shift keying), 16QAM (quadrature amplitude modulation), and 64QAM. The higher order of modulation allows more bits to be encoded per symbol to achieve higher data rate, but it is more prone to interferences (such as 64QAM). However, the lower order of modulation delivers low transmission speed but is more robust against interferences. Table 2.1 shows the bit rates for different modulation schemes under different channel sizes.

2.2.1.4 Duplexing Technologies

802.16 supports both frequency division duplexing (FDD) and time division duplexing (TDD). FDD requires two channels: one for transmission and one for reception while for TDD a single channel is shared by both the uplink and the downlink but separated by different time slots. FDD is designed only for symmetrical traffic with lower spectrum efficiency and higher cost but shorter delay. In contrast, TDD supports both symmetrical and asymmetrical traffic with better frequency usage, but it cannot transmit and receive at the

TABLE 2.1

Bit Rates and Channel Sizes

Channel Size (MHz)	Bit Rate (Mbps)		
	QPSK	16QAM	64QAM
20	32	64	96
25	40	80	120
28	44.8	89.6	134.4

Source: Reprinted with permission from IEEE Standard for Local and Metropolitan Area Networks Part 16: Air Interface for Fixed Broadband Wireless Access Systems, © IEEE 2002.

same time. TDD is more efficient for data transmission while voice traffic can be handled by FDD with minimum delays.

2.2.1.5 Multiplexing Technologies

The multiplexing technologies used in 802.16 are time division multiplexing (TDM)—for downlink channel and time division multiple access (TDMA)—for uplink channel. In TDM, subscribers share the same frequency band but are allocated by different time slots. TDMA is a flexible multiple access scheme in which time slots can be allocated to subscribers according to fixed or contention modes.

2.2.1.6 Quality of Service

To allow quality-of-service (QoS) differentiation, the uplink traffic flows are grouped into four types of applications for 802.16 MAC:

- Unsolicited grant services (UGS): UGS is designed to support constant bit rate services, such as T1/E1 emulation and voice over IP (VoIP) without silence suppression.
- Real-time polling services (rtPS): It is used to support real-time variable bit rate services, such as MPEG video and VoIP with silence suppression.
- Nonreal-time polling services (nrtPS): It is used to support nonreal-time variable bit rate services, such as FTP.
- Best-effort (BE) services: With BE services, packets are forwarded on a first-in-first-out basis using the capacity not used by other services. Web browsing is one example of it.

The 802.16 MAC is connection oriented and every traffic flow is mapped into a connection, which is identified by a CID and assigned to one of the above four service types with a set of QoS and traffic parameters. The UGS traffic flow has the highest priority while the BE service has the lowest.

2.2.2 802.16a

IEEE 802.16a (published in April 2003) is an improved version of 802.16. This standard extends the 802.16 spectrum down to a lower frequency range from 2 to 11 GHz so that it can utilize both the unlicensed and licensed bands and enables NLOS transmission. LOS transmissions are not required in this case because radio waves at 2–11 GHz frequency bands can penetrate into and bend and reflect around buildings and other obstacles to some extent, which are more desirable in urban areas. However, the performance of NLOS is worse than LOS owing to the attenuation when passing through obstacles and the introduction of license-free bands that increase the interference. So, a dynamic frequency selection (DFS) mechanism is specified in 802.16a to reduce such interference. The implementation of DFS enables the mobile device to switch between different radio frequency (RF) channels on the basis of certain channel measurement criteria, such as signal-to-interference ratio. This standard is designed to support a maximum data rate of 75 Mbps at a distance of up to 50 km. In the following sections, the main new features introduced by 802.16a will be discussed.

2.2.2.1 Flexible Bandwidth

A problem existing in the original 802.16 standard is that it is often very difficult for some power-sensitive devices, such as laptops and handheld equipments, to transmit to the BS over long distances if the channel bandwidth is too wide. It is solved in 802.16a by using flexible bandwidth choices, including channel bandwidth between 1.25 and 28 MHz, which provides the flexibility to operate in different frequency bands with varying channel requirements around the world. Because of the interference problem in 2–11 GHz bands, 802.16a systems are more attractive in rural and developing markets where there are sufficient unlicensed spectrums available without interference concerns.

2.2.2.2 Mesh Topology

In addition to PTP and PMP, 802.16a introduces the mesh topology, which is a more flexible, effective, reliable, and portable network architecture based on the multihop concept. Mesh networks are wireless data networks that give the SSs more intelligence than traditional wireless transmitters and receivers. In a PMP network, all the connections must go through the BS, while with mesh topology, every SS can act as an access point and is able to route packets to its neighbors by itself to enlarge the geographical coverage of a network. The architecture of a mesh system is shown in Figure 2.3. The routing across the network can be either proactive (using predetermined routing tables) or reactive (generating routes on demand).

Mesh topology can be divided into two basic categories: switched mesh and routed mesh [11]. In a switched mesh, a fixed route between two network nodes is predetermined and all packets follow the same path during the transmission. If the connection is down or the QoS of the link is degraded, a new route will be established to replace the old one. However, in routed mesh architecture, there is no fixed path from the source to the destination. All packets are forwarded by intelligent network nodes on the basis of the evaluation of link conditions measured by a number of parameters, such as throughput, traffic density, packet loss, interference level, delay, and jitter. Packets from the same source to the same destination may follow different paths and arrive with various delays and jitters. The routed mesh can be further divided into different forms. At one extreme, every node knows all the other nodes in the network, which is called all-knowing mesh. At another extreme, every node only knows its immediate neighbors. An all-knowing

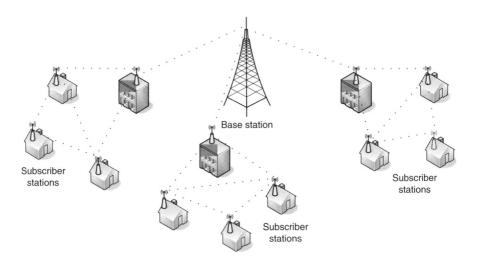

FIGURE 2.3
Mesh network topology.

mesh has better understanding of the whole network and can find the best path for data transmission. However, it is more complicated and expensive owing to the need for large memory size, high processing power, and complex routing algorithm. A trade-off is required to decide the appropriate mesh forms.

Mesh topology is better than its single-hop and directional alternatives. It is more robust against system failure. In a single-hop network, if a single node goes down, so does the whole system. However, in a mesh network, if one node is out of work, the system continues to operate by simply routing packets through an alternative path. Mesh topology can also provide greater redundancy for traffic balancing. In single-hop networks, if too much traffic flows are transmitted simultaneously, a traffic jam may happen and the system will sharply slow down. Mesh networks solve this problem by routing data along an alternative path, where the traffic load is light so that the available bandwidth can be used more efficiently. Another advantage of mesh is the saving of cost. Because network intelligence is distributed to each network node, the number of network management devices, such as BSs, central offices, routers, and switches can be significantly reduced. The backhaul is also no longer needed. In addition, mesh topology can also help the network to adapt to changes and navigate around large obstacles.

However, some problems also arise with mesh topology. The latency increases with the number of network hops, which may degrade the quality of delay-sensitive applications, such as voice traffic. Mesh networks are inherently noisy because wireless mesh links are multidirectional broadcasters that may pick up extra signals. Increasing the number of mesh nodes may also cause scalability issues because the routing tables in them will become more complex.

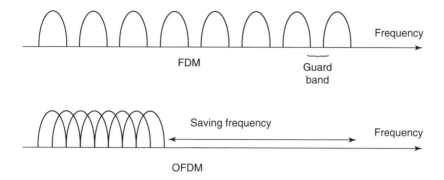

FIGURE 2.4
Comparison between FDM and OFDM in bandwidth utilization.

2.2.2.3 Orthogonal Frequency Division Multiplexing

One important improvement of 802.16a is the usage of orthogonal frequency division multiplexing (OFDM) technology, which allows high-speed bidirectional wireless data transmission in a mobile environment. OFDM is based on the traditional frequency division multiplexing (FDM), which enables simultaneous transmission of multiple signals by separating them into different frequency bands (subcarriers) and sending them in parallel. In FDM, guard bands are needed to reduce the interference between different frequencies, which causes bandwidth wastage (shown in Figure 2.4). Therefore, it is not a spectrum-efficient and cost-effective solution. However, OFDM is a more spectrum-efficient method that removes all the guard bands but keeps the modulated signals orthogonal to mitigate the interference level. As shown in Figure 2.4, the required bandwidth in OFDM is significantly decreased by spacing multiple modulated carriers closer until they are actually overlapping.

OFDM uses fast Fourier transform (FFT) and inverse FFT to convert serial data to multiple channels. The FFT size is 256, which means a total number of 256 subchannels (carriers) are defined for OFDM. In OFDM, the original signal is divided into 256 subcarriers and transmitted in parallel. Therefore, OFDM is referred to as a multicarrier modulation scheme. Compared to single-carrier schemes, OFDM is more robust against multipath propagation delay owing to the use of narrower subcarriers with low bit rates resulting in long symbol periods. A guard time is introduced at each OFDM symbol to further mitigate the effect of multipath delay spread. For more details of OFDM, please refer to Refs. 10 and 12.

2.2.2.4 Adaptive Modulation

Another new feature of 802.16a standard is adaptive modulation, which allows the provision of more flexible services to customers by enabling the BS to dynamically assign modulation schemes to the clients. Like 802.16, 802.16a supports different modulation technologies, including QPSK, 16QAM, and 64QAM. The higher the order of modulation, the higher is the bit rate achieved. However, high-order modulation techniques are more susceptible to interference and noise, which cause higher bit error ratios (BERs). The use of adaptive modulation allows a wireless system to adjust modulation schemes depending on the channel conditions, distance between the BS and the user, weather, signal interference, and other transient factors. In good channel conditions, high-order modulations can be used to increase the data throughput and spectral efficiency. When the radio channel conditions become worse, low-order modulations should be used to maintain a certain BER.

Using adaptive modulation, it is also able to provide a gradation of QoS depending on the distance from the user to the BS. The longer the distance between the BS and the SS, the lower the guarantee of QoS. A BS can choose the highest modulation scheme (64QAM) to increase the throughput of a customer close to it, while the modulation order may be reduced to 16QAM or even QPSK to serve a distant customer. It allows the BS to automatically extend its effective range at the expense of reducing throughput or vice versa. Adaptive modulation maximizes the network performance while ensuring robust RF links in the quickly changing wireless environment.

2.2.3   802.16-2004

IEEE 802.16-2004 is a wireless access technology standard optimized for fixed and nomadic access, which was published in October 2004. It is a combined and improved version of IEEE 802.16, 802.16a, and 802.16c (these three standards are replaced by 802.16-2004 now) in which both the 10–66 GHz and 2–11 GHz frequency bands are specified and the bandwidth can be as narrow as 1.25 MHz. IEEE 802.16-2004 is designed for fixed BWA systems to support multiple services. The goal of this standard is to enable global deployment of innovative, low-cost, and interoperable multivendor BWA products; increase the capacity of competition of BWA systems against their wired counterparts; and facilitate global commercialization of BWA products. IEEE 802.16-2004 does not add any new models in addition to those covered by IEEE 802.16 and 802.16a. Its main features also remain the same and have already been discussed before.

2.2.4   802.16e

All the above standards only focus on fixed broadband systems. However, IEEE 802.16e standard published in February 2006 aims to provide portability and mobility to wireless devices and supports for higher layer handover, which are lacking in the previous standard. 802.16e also enhances the network performance in fixed environment by using orthogonal frequency division multiplexing access (OFDMA). However, the frequency bands suitable for mobility must be below 6 GHz. IEEE 802.16e is also not backward compatible with 802.16-2004 so that hardware/software updates are required to implement it.

Compared with 802.16-2004, 802.16e has lower throughput (up to 15 Mbps), but it supports both hard and soft handoffs. Hard handoffs are based on break-before-make concept, which leads to high latency while soft handoffs use make-before-break approach to minimize the delay. The former is usually used for data transfer, while the latter is more suitable for delay-sensitive applications, such as VoIP and online games.

IEEE 802.16e uses OFDMA to enhance network performance. OFDMA is a multiple-user version of OFDM and is a more flexible way to manage different user devices with various antenna types and form factors. In OFDMA, the whole carrier space is divided into N groups, where each of them includes M carriers. All the carriers are then grouped into M subchannels, each with one carrier per group.

In OFDM, only one user device can use the channel during a single time slot. OFDMA allows multiple users to transmit data simultaneously. A number of users can communicate at the same time using the subchannels allocated to them.

Signal coding, modulation, and amplitude are set separately for each sub-channel based on channel conditions to optimize the utilization of network resources. From the user perspective, subchannelization allows different subchannels to be allocated to different subscribers according to their requirements and channel conditions. One customer can be allocated two or more subchannels. For service providers, subchannelization provides a flexible and efficient bandwidth management solution and a flexible power transmission method. Higher power can be allocated to those subchannels with bad radio conditions.

Using OFDMA, fixed user devices can be supported with the same data rate as OFDM, while mobile users trade off mobility against bandwidth. Compared to OFDM, OFDMA supports larger FFT size of 1024 so that it enables more flexible subcarrier bandwidth allocation [10].

The four main IEEE 802.16 family standards have been introduced in the above sections. Table 2.2 summarizes their main features.

2.2.5 Other IEEE 802.16 Family Standards

In addition to the four main standards discussed before, there are some other IEEE 802.16 family standards that will be briefly introduced in the following sections for completeness. If readers want to learn more about these standards, please refer to Ref. 9.

2.2.5.1 802.16c

It was published in January 2003 as an amendment to 802.16. This standard is aimed to develop the 10–66 GHz BWA system profiles and aid interoperability specifications. It has already been replaced by the IEEE 802.16-2004 standard.

TABLE 2.2

Comparison among IEEE 802.16, 802.16a, 802.16-2004, and 802.16e

	802.16	802.16a	802.16-2004	802.16e
Frequency range	10–66 GHz	2–11 GHz,	2–11 GHz, 10–66 GHz	2–6 GHz
Channel conditions	Line-of-sight only	Nonline-of-sight	Nonline-of-sight	Nonline-of-sight
Channel bandwidth	20, 25, and 28 MHz	1.25–28 MHz	1.25–28 MHz	1.25–20 MHz
Modulation scheme	QPSK, 16QAM, and 64QAM	OFDM, QPSK, 16QAM, and 64QAM	OFDM, QPSK, 16QAM, and 64QAM	OFDM, QPSK, 16QAM, and 64QAM
Network architecture supported	PTP, PMP	PTP, PMP, mesh	PTP, PMP, mesh	PTP, PMP, mesh
Bit rate	32–134 Mbps	Up to 75 Mbps	Up to 75 Mbps	Up to 15 Mbps
Mobility	Fixed	Fixed	Fixed	Pedestrian mobility—regional roaming, maximum mobility support: 125 km/h
Typical cell radius	1–3 miles	Maximum range is 30 miles on the basis of antenna height, antenna gain, and transmit power	Maximum range is 30 miles on the basis of antenna height, antenna gain, and transmit power	1–3 miles
Applications	Replacement of E1/T1 services for enterprises, backhaul for hot spots, residential broadband access, SOHO (small office/home office)	Alternative to E1/T1, DSL, cable backhaul for cellular and WiFi, VoIP, Internet connections	801.16 plus 802.16a applications	802.16-2004 applications plus fixed VoIP, QoS-based applications, and enterprise networking

2.2.5.2   *802.16.2-2001*

It was published in September 2001 and the specification is about recommended practice on coexistence of BWA systems operated in the 10–66 GHz licensed bands. It has been replaced by IEEE 802.16.2-2004.

2.2.5.3   *802.16.2-2004*

IEEE 802.16.2-2004 standard (published in March 2004) recommends for the coexistence of different FBWA systems in both the 10–66 GHz and

2–11 GHz frequency bands and the minimization of interference. In this standard, the coexistence guidelines and criteria, equipment design parameters, system coordination methodology, interference evaluation, and mitigation techniques are recommended to avoid case-by-case coordination.

2.2.5.4 802.16f-2005

This standard was published in September 2005 and is an enhanced version of the IEEE 802.16-2004 standard. The purpose of it is to specify a management information base for the MAC and PHY layers and associated management procedures to enable standardized management of 802.16 devices.

2.2.5.5 IEEE Standard 802.16/Conformance01-2003

This standard was published in August 2003 aiming to evaluate the conformance of a particular implementation. It represents a statement called protocol implementation conformance statement (PICS) to specify which capabilities and options have been implemented and what limitations might prevent interworking for 10–66 GHz BWA systems.

2.2.5.6 IEEE Standard 802.16/Conformance02-2003

This standard was published in February 2004 describing the test suite structure and test purposes for 10–66 GHz BWA systems.

2.2.5.7 IEEE Standard 802.16/Conformance03-2004

This standard was published in June 2004 aiming to specify the conformance and interoperability testing at the 10–66 GHz radio interface.

2.3 WiMAX Applications

WiMAX is a WMAN technology, which fits between WLANs and wireless wide area networks (WANs). It has been developed to provide cost-effective, high-quality, and flexible BWA solutions using certified, compatible, and interoperable equipments from different vendors. WiMAX can provide broadband services to people who could not afford wired broadband services before and cover areas where broadband services have not been available before. Compared to wired broadband technologies, WiMAX has the following advantages: cheaper implementation costs, less monthly ongoing maintenance costs, quicker and easier setup/deployment/reconfiguration/disassembly, less impact on environment, more scalability for future network expanding, and more flexibility.

The distance of a WiMAX connection can be up to 30 miles (50 km) at data rates up to 75 Mbps using both the unlicensed and licensed spectrums. WiMAX has a wide range of applications, from large area coverage to last-mile access and backhauling. In the following sections, we will introduce some of them.

2.3.1   WMANs

One main reason that makes WiMAX popular today is its potential to provide wireless broadband access to metropolitan areas with the same results as traditional MAN technologies but without the difficulty of establishing and marinating the physical transmission medium, such as copper or fiber lines. In metropolitan areas, the existing wired broadband access technologies create a bandwidth bottleneck owing to the contradiction between the limited bandwidth and the high concentration of customers. One way to solve it is to lease high-capacity connections, but it is too expensive to most of the subscribers. Using DSL or cable modem is an affordable solution, but it is difficult and time consuming to implement. The QoS may also be limited by the distance and the quality of wiring. WMANs are supposed to provide wireless broadband services at lower cost but with equivalent or even higher capacity compared to their wired counterparts.

A WMAN based on WiMAX uses PMP architectures to provide broadband services, such as fast Internet and multimedia applications over a radius up to several kilometers. The range of a WMAN network is determined by the available frequency bandwidth, transmit power, and receiver sensitivity. In a WMAN network, each wireless BS is connected with a group of users in an NLOS and a PMP way while the BSs are typically backhauled to the core network via fibers to available fiber nodes or PTP microwave links. Figure 2.5 shows the architecture of a WMAN using WiMAX technology.

2.3.2   Rural Area Broadband Services

A challenge for broadband service providers today is how to deliver services in rural areas. Today, the broadband services in densely populated and business areas are served well by cable and DSL. However, in rural areas, the situation is much worse. Even in many developed countries, the rural users are limited to low-speed dial-up services or without Internet access at all. Because of the huge difference between rural and urban areas, the provision of nationwide broadband services has become a high priority issue for the governments. The challenge now is how to achieve it. Using wired broadband access technologies is obviously not a good solution. The quality of broadband services depends on both the access network and the interconnection between the local access points and the backbone network (known as backhaul). The cost of the backhaul increases with distance. In urban area, it is not a serious problem but in low population density areas, the cost will be unaffordable to many of the users. It is obviously too expensive, too difficult, too remote, and too time consuming to reach these areas using traditional wired technologies. Although satellites can be used to serve these areas, they have disadvantages such as limited upstream bandwidth, spectrum unavailability,

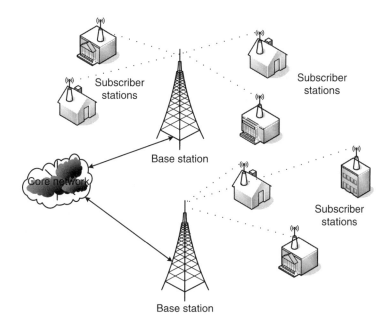

FIGURE 2.5
WiMAX application: WMAN.

and high delay. In this case, WiMAX is the best choice owing to its low cost (for both implementation and ongoing charges) and ease of deployment but with similar performance as DSL. Another benefit of the WiMAX network is its high scalability because the cost of adding a new cell is substantially lower than that in a wired network.

In addition to countryside residential customers, WiMAX technology can also be used for small-to-medium-sized businesses, which are usually located in rural and suburb areas rather than in metropolitan areas, to help them reduce the operation cost.

2.3.3   Last-Mile High-Speed Access to Buildings

The current wired technologies meet some problems while providing last-mile broadband access to buildings, such as high-speed Internet access to residential subscribers, small office/home office (SOHO) users, businesses, campuses, and hospitals. They are expensive owing to the cost of DSL/cable and labor and time-consuming (for example, it takes at least 3 months to install a T1 line in a building) [1]. However, we can use WiMAX instead of wired lines. Compared to its wired counterparts, WiMAX can provide broadband services in a more flexible way (NLOS transmission), at lower cost but with a comparable speed. PMP connections are typically used to link a central location to a group of other locations in this case.

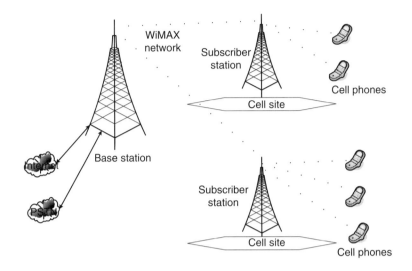

FIGURE 2.6
WiMAX application: Cellular backhaul.

2.3.4 Wireless Backhaul

A backhaul refers to both the connection from the access point back to the provider and the connection from the provider to the core network. Currently, most of the cellular backhaul implementations are done by leasing T1 lines from a third-party service provider, which is expensive. WiMAX can be used as a high-capacity cellular backhaul, which is able to serve multiple cells and expand for future mobile services with lower cost than landline backhaul. With WiMAX, cellular operators can also lessen the dependence on their competitors. Figure 2.6 shows an example of WiMAX backhaul for cellular network.

In addition to cellular systems, WiMAX can also be used as a backhaul for Wi-Fi hot spots, which are growing fast in the world but lack in high-capacity and cost-effective backhaul solutions. WiMAX can combine multiple Wi-Fi hot spots together into a cluster and fill the gap between their coverage areas.

WiMAX backhaul solutions typically use PTP and LOS connections to maximize the effectiveness of the network and can be used in conjunction with other PMP devices to provide a complete solution.

2.3.5 Enterprise/Private Networks

WiMAX can be used by enterprise/private networks to connect remote offices to central offices. The WiMAX enterprise/private networks provide reliable, secure, and high-speed wireless connections between a number of remote buildings and the central office using the PMP topology (shown in Figure 2.7). The WiMAX enterprise/private networks can be used in a wide

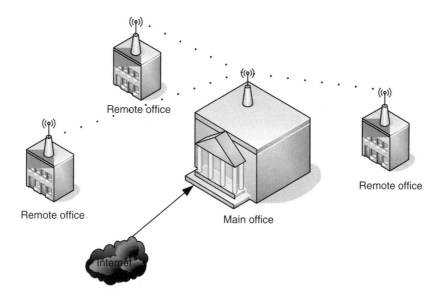

FIGURE 2.7
WiMAX application: Enterprise/private networks.

range of areas including businesses, local governments, education, health, and public organizations.

2.3.6 Wireless Video Surveillance

The increasing demand for video surveillance in high-security, public safety, and crime prevention areas requires a cost-effective, flexible, and reliable monitoring tool. Wireless video surveillance, which combines both the IP and WiMAX technologies, is a solution for monitoring these critical places and is used widely by public and private organizations. Since it can leverage the existing IP network to transfer the security video images via secure and private IP connections and can cover those remote and hard-to-reach locations, wireless video surveillance is typically deployed in shops, retailers, transportation centers, military bases, and parks. Traffic, fire, and flood monitoring are also examples of wireless surveillance applications.

2.3.7 Other Applications

WiMAX can also be used in the following applications:

- Automatic teller machines: WiMAX can help banks to install low-cost ATMs across rural and suburban areas to expand their business.
- Online gaming: WiMAX can be used to provide pervasive and faster online gaming in both the rural and urban areas.

- Multimedia communication: The high capacity of WiMAX enables it to provide multimedia services, such as video conferencing to subscribers in a more flexible and convenient way (access anytime, anywhere).

- Medical applications: WiMAX can be used in medical applications such as remote monitoring patient's vital signs to provide continuous information and immediate response in the event of a patient crisis.

- Vehicular data and voice: WiMAX can help fleet owners, logistic providers, and brokers to locate their vehicles on a real-time basis.

- Sensor networks: Using wireless mesh network technologies WiMAX can create autonomous sensor networks to monitor temperature, air quality, and other factors.

- Backup/redundancy to existing wired networks.

- Telematics and telemetry.

- Mobile transmission of information in emergency situations.

- Real-time monitoring, alerting, and controlling the process of dangerous works.

- Wireless transmission of information of fingerprints, photos, warrants, and other images to and from law-enforcement field personnel.

2.4 Conclusion

It is obvious that WiMAX technology will succeed in a wide range of areas because of its flexibility, interoperability, high-capacity, and low establishment/maintenance cost. Through the efforts of WiMAX Forum, certified WiMAX equipments will provide both the service providers and consumers a convenient, cost-effective, and high-performance alternative to conventional wired access technologies. One of the most exciting aspects of WiMAX is the evolution toward mobility, which is expected to significantly expand its market in the future. Readers can find more detailed knowledge of WiMAX in Refs. 1–6 and 11. Up to date technology development and new applications of WiMAX can be found in Refs. 7 and 8.

References

1. D. Pareek, *The Business of WiMAX*, Chichester, England; Hoboken, NJ: John Wiley, 2006.
2. F. Ohrtman, *WiMAX Handbook: Building 802.16 Wireless Networks*, New York: McGraw-Hill, 2005.

3. D. Sweeney, *WiMAX Operator's Manual: Building 802.16 Wireless Networks*, Berkeley, CA: Apress, 2004.
4. C. Smith and J. Meyer, *3G Wireless with WiMAX and WiFi: 802.16 and 802.11*, New York: McGraw-Hill, 2004.
5. D. Pareek, *WiMAX: Taking Wireless to the MAX*, Boca Raton, FL: Auerbach Publications, 2006.
6. S. Shepard, *WiMAX Crash Course*, New York; London: McGraw-Hill, 2006.
7. WiMAX Forum website, http://www.wimaxforum.org/home/
8. WiMAX.com, http://www.wimax.com/
9. *IEEE 802.16 Published Standards and Drafts*, http://grouper.ieee.org/groups/802/16/published.html
10. G. Nair, J. Chou, T. Madejski, K. Perycz, D. Putzolu, and J. Sydir, IEEE 802.16 medium access control and service provisioning, *Intel Technology Journal*, Vol. 8, No. 3, 2004.
11. A. Ganz, Z. Ganz, and K. Wongthavarawat, *Multimedia Wireless Networks*, Imprint Upper Saddle River, NJ: Prentice Hall PTR, 2004.
12. D. Matiæ, *Introduction to OFDM, II Edition, OFDM as a Possible Modulation Technique for Multimedia Applications in the Range of mm Waves*, http://www.ubicom.tudelft.nl/MMC/Docs/introOFDM.pdf, 1998.

3

WiMAX Technology for Broadband Wireless Communication

Neena Gupta and Gurjit Kaur

CONTENTS

3.1 Introduction

Worldwide interoperability for microwave access (WiMAX), based on the Institution of Electrical & Electronics Engineering (IEEE) 802.16 standards, enables wireless broadband access anywhere, anytime, and on virtually any device. When users want broadband service today, they are generally restricted to a T1, digital subscriber loop (DSL), or cable modem-based connection. However, these wireline infrastructures can be considerably more expensive and time-consuming to deploy than a wireless system. In addition, rural areas and developing countries lack optical fiber or copper wire infrastructure for broadband services, and service providers are unwilling to install the necessary equipments in these areas because of little profit and potential. WiMAX is an ideal technology for backhaul applications because it eliminates expansive leased line or fiber alternative. WiMAX promises to deliver high data rates over large areas to a large number of users. It can provide broadband access to locations in the world's rural and developing areas where broadband is currently unavailable.

WiMAX has numerous advantages, such as improved performance and robustness, end-to-end internet protocol (IP)-based network, secure mobility, and broadband speeds for voice, data, and video. It is a wireless metropolitan area network (WMAN) technology that provides interoperable broadband wireless connectivity to fixed, portable, and nomadic users within 50 km of service area. It allows the users to get broadband connectivity without the need of direct line-of-sight communication to the base station and provides total data rates up to 75 Mbps with sufficient bandwidth to simultaneously support hundreds of residential and business areas with a single base station.

In fact WiMAX is not a technology, but rather a configuration mark, or "stamp of approval" given to equipments that meet certain conformity and interoperability tests for the IEEE 802.16 family of standards. A similar confusion surrounds the term Wi-Fi (wireless fidelity), which like WiMAX, is a certification mark for equipments based on a different set of IEEE standard from the 802.11 working group for wireless local area network (WLAN). Neither WiMAX nor Wi-Fi is a technology but their names have been adopted in popular usage to denote the technologies behind them. This is due to the difficulty of using terms like IEEE 802.11 in common speech and writing. WiMAX is a term coined to describe standard, interoperable implementation of IEEE 802.16 wireless networks in a way similar to Wi-Fi being interoperable of the 802.11 WLAN standards. However, the working of WiMAX is very different from Wi-Fi [1–7].

3.2 WiMAX versus Wi-Fi

The need of network personal computers and other equipments without the cost and complexity of cable infrastructures has brought about rapid growth

TABLE 3.1

The 802.11 Wi-Fi Standards

S. No.	Standard	Frequencies (GHz)	Features
1	802.11a	5	• The modulation technology is OFDM. • Supports speeds up to 54 Mbps.
2	802.11b	2.4	• It uses direct sequence spread spectrum modulation technology. • Supports bandwidth speeds up to 11 Mbps.
3	802.11g (Approved in June 2003)	2.4	• The modulation technology is OFDM. • Supports speeds up to 54 Mbps.

in the Wi-Fi market over the past 6 years. Wi-Fi is a local area-networking standard developed by the IEEE 802.11 working group. Various 802.11 Wi-Fi standards have been tabulated in Table 3.1. It is used for close-range indoor applications and for Internet accessing of a bunch of computers in a home or an office. In contrast, WiMAX is an 802.16 standard-based technology for a last-mile wireless broadband. In Wi-Fi, devices are omnidirectional, finding access points wherever they are; while in WiMAX, devices face an access point, usually called a base station. Users of Wi-Fi devices are expected to hear each other and defer transmission if the network is busy, while in WiMAX, users transmit only when instructed by the base station. External modification to the standards through hardware and software allows Wi-Fi products to become a metro-access deployment option. Each standard uses a different frequency and radio modulation technology [9].

The 802.11 standard has the provision for 64 subcarriers. These individual carriers are sent from the base station to the subscriber station or client and are then reconstituted at the client side. But in nonline-of-sight situation these carriers hit trees, buildings walls, and other objects, which in turn reflect the signal and create multipath interference. These factors were taken into consideration while developing the 802.16-2004 standard.

The 802.11 standard uses a carrier sense multiple access/collision avoidance protocol that listens to the network to avoid transmission collision. This protocol broadcasts a signal onto the network to know about the collision scenarios, and if there is a chance of collision, it informs all other devices to stop the broadcast; whereas WiMAX uses a scheduling protocol, with all scheduling owned by the base station, thus improving reliability of the system [5].

In Wi-Fi networks, as the number of users increases, the efficiency of the network decreases; while in 802.11 g standard, a single user can access 30 Mbps bandwidth, but as the number of users increases, the per user throughput decreases. Also the range of the Wi-Fi system is not very large. Generally, omnidirectional antennas are used for line-of-sight communication. These antennas send signals in all directions as shown in Figure 3.1. So by using them some power is wasted. Directional antennas can be used to save the power as well as to increase the range. These antennas have much higher

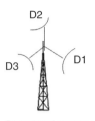

Omnidirectional
antenna

Directional antennas

FIGURE 3.1
Antenna used in Wi-Fi networks.

gain than omnidirectional antennas. Although they send signal in a particular direction, the number of antennas required is more for the same coverage.

Wi-Fi works in unlicensed spectrum using the 2.4 and 5 GHz bands. Wi-Fi is a cheap and easy way of providing local connectivity at high speed. WiMAX uses licensed spectrum and has strong authentication mechanisms built in. It has considerably greater range than Wi-Fi. When taken together WiMAX and Wi-Fi are complementary to each other than competitive.

WiMAX is referred to as "Wi-Fi on steroids." It has the potential to enable millions to access the Internet wirelessly, cheaply, and easily. WiMAX wireless coverage is measured in square kilometers/miles, while in case of Wi-Fi it is measured in square meters/yards. WiMAX base station can beam high-speed Internet connections to homes and businesses in a radius of up to 50 km (31 miles). These base stations can cover an entire metropolitan area, transforming that area into a WMAN and allowing true wireless mobility within it, as opposed to hot spot hopping by Wi-Fi.

WiMAX standard has the spectrum range from 2 to 11 GHz. The WiMAX specification improves upon many of the limitations of the Wi-Fi standard by providing increased bandwidth and stronger encryption. The standards of WiMAX are given in Table 3.2.

This technology supports 70 Mbit/s of shared data rate. According to properties it has enough bandwidth to simultaneously support more than 60 businesses with T1-type connectivity and well over a thousand homes at 1 Mbit/s DSL level connectivity [16].

3.3 WiMAX Architecture

Wireless broadband systems have been in use for several years, but the deployment of this new standard, that is, WiMAX marks the maturation of the industry and brings in a new level of competitiveness for nonline-of-sight

TABLE 3.2

WiMAX IEEE 802.16

Standard	Frequency Range	Features
IEEE 802.16a Jan 2003	Having licensed and license-exempt frequencies (2–11 GHz)	• At lower frequencies the signals can penetrate barriers and do not require line of sight between transmitter and receiver.
IEEE 802.16b	5–6 GHz	• Provides high quality of service for transmission of real-time voice and video.
IEEE 802.16c	10–66 GHz	• This encourages more consistent implementation and more interoperability.
IEEE 802.16d or 802.16-2004 (June)	2–11 GHz fixed	• Adds support to 802.16a for indoor customer premise equipment. • This standard combines the physical (PHY) layer and media access controller (MAC) layer, ensuring a uniform base for all WiMAX stations. • It uses mounted antenna at the subscriber site. • Uses orthogonal frequency division multiple access (OFDMA) for optimization of wireless data services.
IEEE 802.16e	2–6 GHz portable	• Adds support for mobility. • Using OFDMA, it divides the carriers into multiple subscribers. It goes a step further by then grouping multiple subscribers into subchannels.

wireless broadband services. The basic WiMAX architecture is shown in Figure 3.2.

The network architecture consists of a base station in the center of the city, with the base station communicating with all the substations or access points. Each sector can provide broadband connectivity to dozens of businesses and hundreds of homes. WiMAX can further be connected to one or more Wi-Fi access points to connect with a Wi-Fi enabled Laptop, or a standard Ethernet cable attached to a computer or LAN [18,19].

The various parameters of IEEE 802.16 standard in WiMAX are related to the MAC and PHY layers. To ensure that resulting 802.16-based devices are in fact interoperable, an industry consortium called the WiMAX Forum was created. The WiMAX Forum develops guidelines known as profiles, which specify the frequency band of operation, the physical features to be used, and a number of other parameters. The WiMAX Forum has identified several frequency bands for the initial 802.16d products, in both licensed (2.5–2.69 GHz and 3.4–3.6 GHz) and unlicensed spectrums (5.725–5.850 GHz). The IEEE

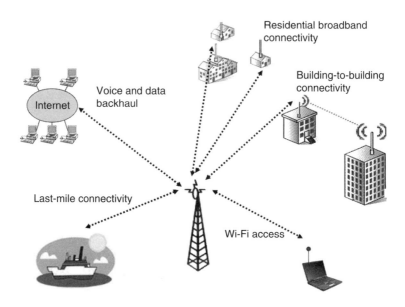

FIGURE 3.2
WiMAX architecture.

802.16a/d standard defines three different PHY layers that can be used in conjunction with the MAC layer to provide a reliable end-to-end link [17].
   The air interface specifications are as follows:

1. WMAN-SCa: A single carrier modulated air interface.
2. WMAN-OFDM: It is a 256 carrier orthogonal frequency division multiplexing scheme. It uses the time division multiple access (TDMA) technology.
3. WMAN-OFDMA: It is a 2048 carrier OFDM scheme. Multiple access is provided by assigning a subset of the carriers to an individual receiver. This is also referred to as orthogonal frequency division multiple access (OFDMA). The OFDMA-based systems are more suitable for nonline-of-sight operation. The WiMAX architecture is based on a packet switched framework, including native procedures based on the IEEE 802.16 standard and its amendments. It allows modularity and flexibility to accommodate a broad range of deployment options such as licensed or license-exempt frequency bands, co-existence of fixed, nomadic, portable, and mobile usage models, etc [6].

   To ensure global implementation, WiMAX can use variable channel bandwidth. The channel bandwidth can be an integral multiple of 12.5, 1.5, and 1.75 MHz with the maximum of 20 MHz. The bandwidth request and grant mechanism have been designed to be scalable, efficient, and self-correcting. The 802.16 access system does not lose efficiency when presented with multiple connections per terminal, multiple quality of service (QoS) levels per terminal, and a large number of statistically multiplexed users. The MAC layer is divided into convergence-specific common part sublayers. These sublayers are used to map the transport layer specific to a MAC. It is flexible enough to efficiently carry any type of traffic. The common part sublayer is independent of the transport mechanism and responsible for fragmentation and segmentation of MAC service data units into MAC protocol data units, QoS control and scheduling, and retransmission of MAC protocol data units.

The nonline-of-sight technology and enhanced features in WiMAX make it possible to use an indoor customer premise equipment (CPE). But it has two main challenges:

1. Overcoming the building penetration losses
2. Covering reasonable distances with lower transmit powers and antenna gain that are usually associated with indoor CPEs

The above problems can be solved by using OFDMA technology, subchannelization, adaptive modulation, error-correcting techniques, directional antennas, or power control [10–12].

3.3.1 OFDMA Technology

OFDMA technology provides a solution to overcome the challenges of nonline-of-sight communication. OFDMA gives more flexibility to 802.16 profiles when managing different user devices with a variety of antenna types and form factor. It helps to reduce interference for user devices with omnidirectional antennas. The WiMAX OFDMA waveform can operate with the larger delay spread of the nonline-of-sight environment. Because of the use of cyclic prefix and OFDM symbol rate as shown in Figure 3.3, the OFDMA waveform eliminates the intersymbol interference and complexities of adaptive equalization.

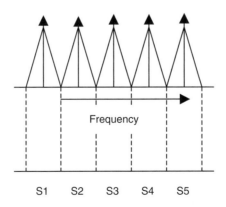

FIGURE 3.3
Orthogonal frequency division multiple access.

Basically, in OFDMA systems the subchannels maintain their orthogonality in a multipath channel. The number of multipath components does not limit the performance of the system as long as the multipaths are within the cyclic prefix. So this type of systems are robust to multipath effects. The subchannel orthogonality within the cyclic prefix window also relaxes the time synchronization requirement. In OFDMA, users are allocated different portions of the channel; there is very less multiple access interference between multiple users. So OFDMA can support higher order uplink modulations and achieve higher uplink spectral efficiency.

3.3.2   Subchannelization

Subchannelization defines subchannels that can be allocated to different subscribers depending upon the channel conditions and their data requirements. It concentrates the transmit power into fewer OFDM carriers and increases the system gain that can be used to extend the range of the system, to reduce the power consumption, and to overcome the building penetration losses. It gives more flexibility in managing the bandwidth and power transmission.

3.3.3   Adaptive Modulation and Coding

The 802.16 standard defines several combinations of modulation and coding rate that can be used to achieve the trade-off of data rate, robustness, and interference conditions. In one combination 802.16 standard can use Reed–Solomon code as the outer code and block codes as the inner code. This interleaving is employed to reduce the effect of burst errors. Turbo codes can improve the coverage and capacity of the system at the cost of complexity of the system. The 802.16 standard can use binary phase shift keying (BPSK), 16 quadrature amplitude modulation (QAM), and 64 QAM. In each data stream a total of eight pilot subcarriers are inserted to constitute the OFDM symbol.

The system can adjust the modulation scheme depending upon the requirement. If the quality of the signal is high then the highest modulation scheme can be used, which can give the system a large capacity. When the quality of the signal is poor, it shifts to the lower modulation scheme so that connectivity can be maintained properly. Owing to this type of modulation scheme the range can also be increased with the help of a high modulation scheme. The problem of frequency-selective fading or burst errors can be reduced by using proper error-correcting techniques.

3.3.4   WiMAX Security

WiMAX security is an important issue while determining the performance of the system. Security is probably a good factor to explain the difference between the robust base station of WiMAX and the ways in which individual vendors can still differentiate their products beyond the features that the base standard offers. Access and authentication remain key wireless concerns for enterprise buyers and users. The main factor for week security can be the insecure coding at the software driver level, which can be exploited by clever hackers. But the new coding methods, such as Reed–Solomon, convolutional codes, block codes, and other interleaving code methods, can detect and correct errors [13].

Typical residential service does not require the kind of security a bank, a hospital, or a government often needs. But WiMAX can handle it. Support exists for mutual device/user authentication, flexible key management plane message protection, and security protocol optimization for fast handovers.

In WiMAX, for traffic encryption advanced encryption standard-counter with CBC-MAC (AES-CCM) ciphers are used. This cipher protects all the data over the MAC layer. WiMAX supports fast handover. For this a three-way handshake scheme is used to optimize the reauthentication mechanism. With soft handoff, which is typically employed in voice-centric mobile networks, multiple base stations, in a mobile active set of base stations, transmit the same data simultaneously to minimize the handoff delay. Soft handoff, however, is not a good approach since it is neither spectrally efficient nor necessary for delay-tolerant data traffic.

WiMAX supports "network-optimized hard handoff" for bandwidth-efficient handoff with reduced delay, achieving a handoff delay of less than 50 ms. WiMAX also supports fast base station switch and macro diversity handover as an option to further reduce the handoff delay.

A consistent and extensible authentication framework is deployed in WiMAX for authentication mechanisms in home and operator network scenarios. WiMAX uses standard secure IP address management mechanisms between the mobile station and its home. In WiMAX, all traffic is encrypted with CCMP (i.e., counter mode with cipher block chaining message authentication code protocol). CCMP uses AES to provide the encryption for secure transmission as well as data authentication for data integrity. For end-to-end authentication, WiMAX uses extensible authentication protocol, which relies on the transport layer security (TLS) standard that uses public key cryptography [14,15,20].

3.3.5 WiMAX Advantages

The IEEE 802.16 standard is designed for WMAN networks. It provides interoperable broadband wireless connectivity to fixed and nomadic users. It provides up to 50 km of service area and allows the users to get broadband connectivity without the need of direct line of sight to the base station. It provides a total data rate of up to 75 Mbps, which is enough to simultaneously support a lot of business and home requirements. The advantages of WiMAX are given as follows.

3.3.5.1 *High Capacity*

A single WiMAX main station can serve hundreds of users. It targets a range of up to 31 miles with target transmission rate exceeding 100 Mbps. By using higher modulation, bandwidth can further be increased. Through WiMAX one can transfer data, voice, Internet, video images, pictures, video conferencing, etc., at a very high data rate. So WiMAX can provide sufficient bandwidth to the end users.

3.3.5.2   Quality of Service

The MAC layer of the WiMAX architecture is responsible for Qos. Subchannelization and different coding schemes enable end-to-end QoS. High data rate and flexible scheduling can enhance the QoS.

3.3.5.3   Flexible Architecture

The architecture of WiMAX is highly flexible. Depending upon the requirement it can connect different stations on point-to-point or point-to-multipoint basis. Further the range can be increased with the help of directional antennas.

3.3.5.4   Mobility

In WiMAX, the user device can maintain an operating network data service session for real-time application as it moves at vehicular speeds within the network coverage area. It supports optimized handover schemes with latencies less than 50 ms to ensure real-time application such as voice over Internet protocol (VoIP) without service degradation. Flexible key management assures that security is maintained during handover.

3.3.5.5   Improved User Connectivity

The IEEE 802.16 standard keeps more users connected by virtue of its flexible channel bandwidths and adaptive modulation. WiMAX uses channels narrower than the fixed 20 MHz channels used in Wi-Fi. It can serve lower data rate users without wasting bandwidth. Adaptive modulation helps to connect them in the noisy or low-signal strength conditions.

3.3.5.6   Robust Carrier Class Operation

As the number of users accessing the data increases, the aggregate bandwidth is shared because of which the individual throughput starts decreasing linearly. The decrease is lesser than what is experienced under Wi-Fi. So this standard is designed for carrier class operation.

3.3.5.7   Scalability

WiMAX system offers scalability in network architecture as well as in radio access technology. It provides a great deal of flexibility in network deployment options and service offerings. It is designed to work in different forms of channelization from 1.25 to 20 MHz to comply with varied worldwide requirements. It can also fulfill the needs such as providing affordable Internet access in rural areas versus enhancing the capacity of broadband access in metro and suburban areas only.

3.3.5.8 Nonline-of-Sight Connectivity

WiMAX is based on OFDM technology and can handle nonline-of-sight connectivity. This capability helps WiMAX to communicate in a nonline-of-sight environment, which other wireless products cannot. The nonline-of-sight coverage can further be increased by using directional antennas or adaptive modulation [23].

3.3.5.9 Cost Effectiveness

Mass adoption of the standard and the use of low-cost, mass-produced chipsets can reduce costs dramatically, and the resultant competitive pricing will provide considerable cost saving for service providers and end users. Further, base stations and base station equipments need not be installed in totality at the outlet, but can be deployed over a period of time to address specific market segments or geographical areas of Internet to the operator.

3.3.5.10 Fixed and Nomadic Access

WiMAX can provide both fixed and nomadic access to its users. In fixed access, the user device is assumed to be fixed in a single geographical area for the duration of the network subscription. Here the user can connect and disconnect from the network. It can select the best base station while entering the network. The user is associated only with the same base station sector or cell, and any reassociation with other cell is controlled by the network.

In nomadic access, the user device is assumed to be fixed in a geographical location at least as long as the network data service session is in operation if the user shifts to a new location in the same wireless network. The user subscription is recognized, and a new data service session is established. The user device is associated with the same base station during a data service session. So WiMAX complements third-generation mobile networks by providing "nomadic" broadband access. Vendors can now compete to sell their equipment, which benefits the customer base by providing lower costs and enabling broadband access in emerging markets [21,23].

3.3.6 WiMAX Applications

WiMAX attribute opens the technology to a wide variety of applications because of its high transmission rate and large range. It serves as a backbone for Wi-Fi for connectivity to the Internet. It can provide broadband connectivity over large coverage area as compared to 802.11 standard. WiMAX is a broadband wireless communication system, which enables convergence of mobile and fixed broadband networks through a common wide-area and flexible network architecture. The mobile WiMAX air interfaces use OFDMA for improvement in multiple path interference in nonline-of-sight environment. Its ability to support both line-of-sight and nonline-of-sight connections makes it suitable for ubiquitous services offered in rural and urban areas alike. High speed and symmetrical bandwidth satisfy the needs of individual customers, public administration, and enterprises of all sizes [8].

The technology also provides fast and cheap broadband access to markets that lack infrastructure (fiber optics or copper wire), such as rural and unwired countries. Currently, several companies offer proprietary solutions for wireless broadband access, many of which are expensive because they use chipsets from adjacent technologies, such as 802.11. Early field experiments in various countries confirm that expectations in terms of coverage, performance, and usage scenarios are indeed justified. WiMAX has changed the scenario of wireless broadband from proprietary solutions to a standards-based industry. It supports fast Internet access, high-quality audio and video communications, education, entertainment, telemedicine, telemetering, and telesurveillance.

WiMAX supports personal broadband services on both fixed and mobile settings because of its high spectral efficiency and wide channelization as well as the advanced antenna technologies. This flexibility in providing both fixed and mobile access within the same infrastructure is unprecedented among wireless technologies, which are typically optimized for either mobile or fixed access [16].

3.3.6.1   Cellular Application

The main merit of WiMAX is in the area of mobile service. For a large number of cell phone operators the major monthly operating expense on T1 backhaul that supports their base stations as shown in Figure 3.4. A WiMAX substitute for the cell phone infrastructure could be operated with as little as 10% of T1 backhaul. While replacing a cell phone infrastructure with WiMAX one can send a large amount of data because the bandwidth of WiMAX is far greater. The data can include voice, mobile data, TV, videoconferencing, video on demand, etc.

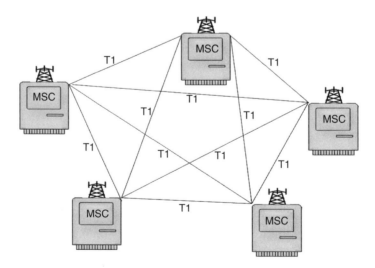

FIGURE 3.4
Cellular architecture using T1 as backhaul.

3.3.6.2 WiMAX Military Applications

As WiMAX uses higher frequencies than current military and commercial communications, existing antenna towers share a WiMAX cell tower without compromising the current communication services. WiMAX can be used to support training and war game simulations. An initial deployment of WiMAX has already been constructed by the U.S. Army Fortdix. The U.S. army is testing prestandard WiMAX gear and Xacta secure wireless system from Telos Corporation in Fort Carson in Colorado for point-to-point and point-to-multipoint communications.

The forces at different locations can be connected through WiMAX as shown in Figure 3.5. They can exchange their information from multiple sources, rapidly and flexibly. This is ideally suited to meet the demands of the tactical defense operations model. The mobile antennas can be attached to a vehicle and the latest data can be provided to the soldiers. A communication from command centers can be made to the different centers, regardless of the distance, and directions can be delivered to the army people. The best part of WiMAX is the handover strategy. It uses "make-before-break" sequence rather than "break-before-make" sequence.

3.3.6.3 Medical Applications

In an emergency situation where patients require immediate medical support, WiMAX can serve as the foundation of a mobile hospital. It can be a platform for e-health. In e-health services a doctor can diagnose his patient at some

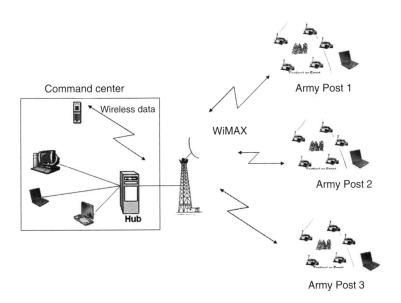

FIGURE 3.5
WiMAX architecture for military applications.

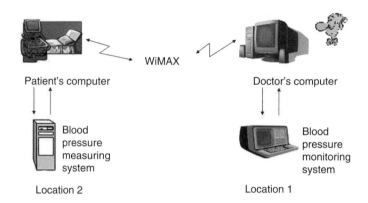

FIGURE 3.6
Medical applications.

far location with the help of e-media. The doctor's computer equipped with the medical instruments can be connected to the patient's computer through WiMAX.

A patient at location 2 can send his reports, for example, blood pressure, through his computer to the doctor's computer as shown in Figure 3.6. The doctor can diagnose the patient's disease and give him necessary treatment. The connection between the doctor and the patient is through the Internet. The two computers are connected through WiMAX.

Also in some emergency situations, a video consultation with a doctor can be set up and the doctor can instruct the paramedic to mobilize the victim without inflicting further damage. With WiMAX, mobile hospital vans can communicate data and other instructions within a disaster zone. The information through WiMAX can be encrypted and made secure. So in diverse conditions WiMAX can provide to the patient valuable information recommended by doctors over large distances [2].

3.3.6.4   Security Systems

WiMAX offers a simple and convenient system for security on the borders and within the country to save the nation from some terrorist attacks.

A video camera can be mounted on WiMAX antenna or some separate pole, which can be controlled at the headquarters as shown in Figure 3.7. This camera will keep an eye over the different activities of the enemies thereby assisting in security planning. It can also be used to provide video surveillance of smuggling and illegal entries along the borders.

WiMAX is a medium for the security of not only army but also navy. Through the use of WiMAX one can monitor the activities on the sea. A video camera that is mounted on the antenna of a shipyard can monitor the nearby activities and report to the headquarter as shown in Figure 3.8. So WiMAX can effectively monitor shipyards, nuclear facilities, and key transport routes.

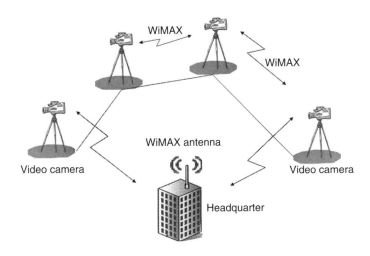

FIGURE 3.7
WiMAX architecture for security applications.

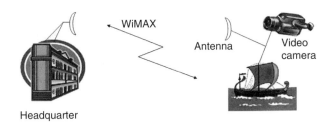

FIGURE 3.8
WiMAX architecture for surveillance.

Video surveillance application through WiMAX can give a platform to the government to improve the security of the nation.

3.3.6.5 Disaster Application

WiMAX can be used in recovery from disasters, such as earthquakes and floods, when the wired networks break down. It helps in connecting the disaster location to telephone services, hospitals, and other important services. In recent hurrican disasters, WiMAX networks were installed to help recovery missions. WiMAX can enable efficient communications with emergency operation centers regardless of the distance. Similarly, WiMAX is used as backup links for broken wired links.

3.3.6.6 Connectivity of Banking Networks

The banking system where security is the major concern can be connected through the WiMAX networks. Owing to the broad coverage and large connectivity, WiMAX can connect a large number of diversely located banks and

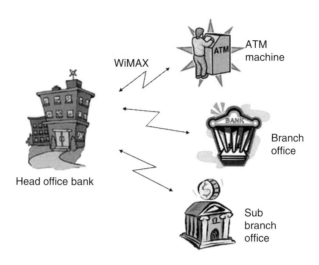

FIGURE 3.9
WiMAX connectivity for banking system.

ATM locations as shown in Figure 3.9. WiMAX networks provide not only security but also a high degree of scalability. Through WiMAX, telephone voice, financial transactions, email, Internet, intranet, surveillance, and close circuit television (CCTV) type of data can be communicated easily.

3.3.6.7   Public Safety

Through WiMAX, public safety agencies can be connected with each other. During any mishap, such as accident, fire, etc., the control office can send its command to the police station, hospital, or fire brigade office as shown in Figure 3.10. The corresponding agencies immediately can connect to the accidental location by using WiMAX-enabled vehicles.

The video images and data from the site of accidental location can be sent to corresponding agencies. These data can be examined by the experts of the emergency staff and accordingly prescription can be communicated. A video camera in the ambulance can send the latest images of the patient before the ambulance reaches the hospital so that the doctors can get ready for further action quickly. Through WiMAX, a fireman can download the data about the best route to a fire scene.

3.3.6.8   Campus Connectivity

Campus system requires high data capacity, a large coverage, and high security. WiMAX can connect various blocks within the campus.

Through this connectivity voice, data, and video information can be sent to various interconnecting blocks as shown in Figure 3.11. It is very difficult to connect various blocks through cables because the lead time to deploy a wired solution is much longer than the lead time to deploy a WiMAX solution. This

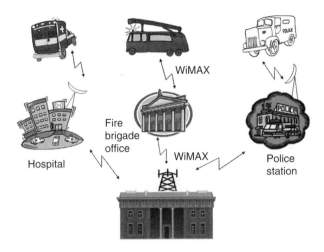

FIGURE 3.10
WiMAX connectivity for public safety.

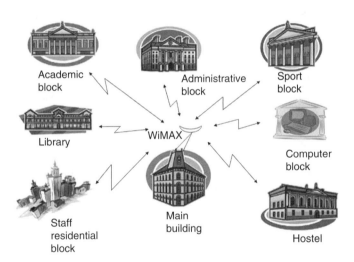

FIGURE 3.11
WiMAX campus connectivity.

connectivity not only reduces the paper work circulation but also ensures fast data transfers.

3.3.6.9 Educational Building's Connectivity

WiMAX can connect boards, colleges, schools, and the main head offices as shown in Figure 3.12. Through this, telephone voice, data, email, Internet, question papers, intranet, video lectures, presentations, and students' results can be communicated at a very high rate.

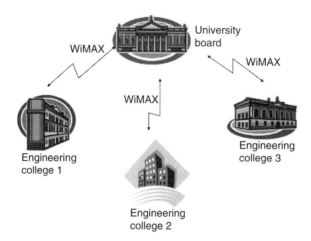

FIGURE 3.12
WiMAX educational building connectivity.

By video conferencing the students can interact with the teachers of another institution (i.e., engineering college, medical colleges, etc.) as shown in Figure 3.12. A camera at college 1 delivers real-time classroom instruction to college 2, allowing the colleges to simultaneously deliver instruction from a recognized subject matter expert to a large number of students. Colleges and schools in rural areas can be connected through WiMAX with other institutions having better facilities through WiMAX so that remotely located students can also be benefitted.

Hence it can be concluded that this broadband wireless standard supports both the computer and telecom industries worldwide, making this technology highly cost effective. It helps enterprises, consumers, public services, and people in urban and rural areas over a large range with high data throughput [22].

References

1. *Beyond 3G? Personal Broadband*, by Monica Paolini, Senza Fili Consulting, August 2006, White Paper The Emergence of WiMAX.
2. *WiMAX for Government Grade Secure Mobility*, NORTEL Government Solutions, www.nortelgov.com
3. *Demystifying WiMAX*, Pyramid research, Global/Business strategies Group, December 1, 2003.
4. *Mobile WiMAX: A Performance and Comparative Summary*, in www.wimaxforum. org
5. *Understanding Wifi and WiMAX com as Metro – Access Solution*, White Paper, Intel.
6. *A Long Road to WiMAX*, IEE Review, October 2005, www.iee.org/review, pp. 32–42.

7. *WiMAX versus 3G Threat or Opportunity*, IEE Communications Engineer, December/January 2005/06, pp. 38–40.
8. *WiMAX Enabled all Electric Sports Car*, IEE Communications Engineers, December/January 2005/06, pp. 41.
9. *WiMAX and Wifi: Separate and Unequal*, by Steven M. Cherry, IEEE Spectrum, March 2004, pp. 16.
10. *Achieving Wireless Broadband with WiMAX*, Steven J. Vanghan-Nichols, Computer, Industry Trends, 2004, pp. 10–13.
11. *WiMAX is coming*, IEE Communications Engineer, August/September 2004, www.ieeorg/communications/magazine
12. *WiMAX in Depth*, by Paul Piggin, IEE Communications Engineer, October/November 2004, pp. 37–39.
13. *INTEL Consigns WiMAX to Science History*, IEE Review, April 2005, 12 pp, www.iee.org/review.
14. *What Will it Take to Move from 802.16 to 802.16E*, by Rupert Baines, IEE Communications Engineer, August/September 2005, pp. 30–31, www.ieee.org/communications
15. *International Telecoms Synchronization Forum and Workshop*, IEE Communications Engineer, August/September 2006, pp. 32–34.
16. *WiMAX: The Emergence of Wireless Broadband*, by Zakhia Abichar, Yanlin Peng, and J. Morris Chang, IT Professional, IEEE Computer Society, July/August 2006.
17. *Broadband Wireless Access With WiMAX/802.16: Current Performance Benchmarks & Future Potential*, by Arunabha Ghosh and David R. Wolter, IEEE Communications Magazine, February 2005, pp. 129–136.
18. *Impact of Wireless (Wifi & WiMAX) on 3G and Next Generation—An Initial Assessment*, by Fauzi Behmann, Freescale Semiconductor.
19. *Fixed, Nomadic, Portable and Mobile Applications for 802.16.2004 and 802.16e WiMAX Networks*, by Senza Fili, www.wimaxforum.org
20. *Mobile WiMAX: A Technical Overview and Performance Evaluation*, by www.wimaxforum.org
21. *Mobile WiMAX—Part 1: A Comparative Analysis*, www.wimaxforum.org
22. *Can WiMAX Address Our Applications?* by Westech Communications Inc., October 24, 2005, www.wimaxforum.org
23. *WiMAX Technology, LOS & NLOS Environment*, White Paper, ISSUE-1 SR Telecom.

4

VoIP over WiMAX

Mainak Chatterjee and Shamik Sengupta

CONTENTS

In spite of the growing popularity of data services, voice services still remain the major revenue earner for the network service providers. The two most popular ways of providing voice services are the packet switched telephone networks (PSTN) and the wireless cellular networks. Deployment of both these forms of networks require infrastructures that are usually very expensive. Alternative solutions are being sought that can deliver good quality voice services at a relatively lower cost. One way to achieve low cost is to use the already existing IP infrastructure. Protocols used to carry voice signals over the IP network are commonly referred to as voice over IP (VoIP) protocols. In more common terms, it signifies the phone service (voice) over the Internet using IP.

There are two major reasons behind the recent thrust for VoIP service. First, VoIP services induce lower cost than traditional voice services. This is mainly due to the existing network infrastructure and underutilized network capacity. Second reason behind the popularity of VoIP service is its increased functionality. Incoming phone calls are automatically routed to a VoIP phone wherever it is plugged and in the most extreme case, users see VoIP phone calls as free (at the expense of Internet service).

Though many delay-sensitive applications are supported over the IP network, supporting real-time applications, such as VoIP, has many challenges [1,2]. VoIP requires minimum service guarantees that go beyond the best-effort structure of today's IP networks. Though some codecs are capable of some level of adaptation and error concealment, VoIP quality remains sensitive to performance degradation in the network. Sustaining good quality VoIP calls becomes even more challenging when the IP network is extended to the wireless domain—either through 802.11-based wireless LANs (WLANs) or third-generation (3G) cellular networks. Such a wireless extension of services is becoming more essential as there is already a huge demand for real-time services over wireless networks. Though, bare basic versions of services such as real-time news, streaming audio, and video on demand are currently being supported, the widespread use and bandwidth demands of these multimedia applications are far exceeding the capacity of current 3G and WLAN technologies. Moreover, most access technologies do not have the option to differentiate specific application demands or user needs. With the rapid growth of wireless technologies, the task of providing broadband *last-mile* connectivity is still a challenge. The last mile is generally referred to as a connection from a service provider's network to the end—user either a residential home or a business facility. Such wireless solutions avoid the prohibitive cost of wiring homes and businesses and allow a relatively faster deployment process. Among the emerging wireless broadband access technologies, WiMAX (worldwide interoperability of microwave access) is perhaps the strongest contender that is being supported and developed by a consortium of companies [3].

4.1 WiMAX—The IEEE 802.16 System

WiMAX is a wireless metropolitan access network (MAN) technology that is based on the standards defined in the IEEE 802.16 specification. This standard-based approach is not only simplifying but also unifying development and deployment of WiMAX. WiMAX is envisioned as a solution to the outdoor broadband wireless access that is capable of delivering very high-speed data up to a range of 30 miles, thus, posing a strong competition to the existing last-mile broadband access technologies such as cable and DSL.

4.1.1 Core Components and Topology

The core components of a WiMAX system are the base stations (BSs) and the subscriber stations (SSs) otherwise known as the CPEs (consumer premise equipments) as shown in Figure 4.1. The BS transmits in the downstream direction to the various SSs, which in turn respond to the BS in the upstream direction.

The 802.16 standard can be used in a point-to-point (PTP), point-to-multipoint (PMP), and mesh topology modes. The effective range of the BS can be increased by using omnidirectional or directional antennas.

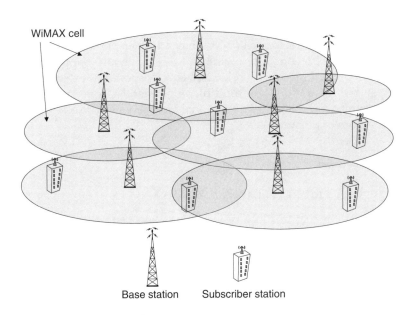

FIGURE 4.1
802.16 topology.

An 802.16-based system often uses fixed antenna at the SS site, usually mounted on the roof. A fixed SS typically uses sectored/directional antenna while a mobile or portable SS usually uses an omnidirectional antenna. The BS controls activity within the cell, including access to the medium by SSs, allocations to achieve quality of service (QoS), and admission to the network on the basis of network security mechanisms.

4.1.2 Frequency of Operation and Service Capacity

The initial 802.16 standard recommended transmission at 10–66 GHz requiring line of sight (LOS); thus multipath is negligible. As a result, in rural areas or in areas with high LOS, this initial standard is effective. However in urban areas, where multipath is inevitable, 802.16 system operating at 10–66 GHz might not be very effective. To address this issue, a new amendment 802.16a [4] has been standardized that overcomes the difficulties of the original 802.16 standard [10]. 802.16a operates in the 2–11 GHz spectrum. Owing to longer wavelength, LOS is not necessary and multipath is significant.

WiMAX uses multiple channels for a single transmission and can provide bandwidth of up to 350 Mbps [5]. The use of orthogonal frequency division multiplexing (OFDM) increases the bandwidth and data capacity by spacing channels very close to each other but still manages to avoid interference because of orthogonal channels. A typical WiMAX BS provides enough bandwidth to cater to the demands of more than 50 businesses with T1 (1.544 Mbps) level services and hundreds of homes with high-speed Internet access. For residential broadband access, WiMAX has a higher potential compared to 802.11-based Wi-Fi technology owing to both range and bandwidth. Even though Wi-Fi-based mesh networks are being proposed to extend coverage, performance degradation with multiple hops is still a concern. A comparison of coverage and data rates of personal area network (PAN), WLAN and WiMAX is shown in Figure 4.2.

4.2 The PHY and MAC Layer of WiMAX

Let us discuss the features of physical (PHY) layer and the medium access control (MAC) layer of WiMAX as that would help us to understand the remainder of this chapter.

4.2.1 The PHY

The distinctive and most critical requirement of the 802.16 PHY is that it has to provide high performance while keeping the complexity low. Applications like VoIP require flexibility for downstream transmission with support for a number of users with possible variable throughputs. 802.16 also supports multiple access on the upstream transmission. Multiple carrier modulation is

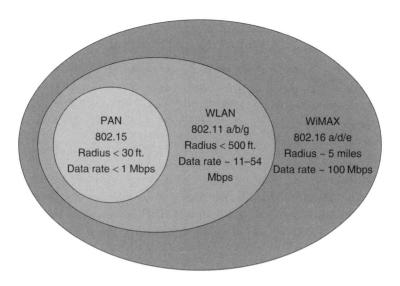

FIGURE 4.2
802.16 system compared to other IEEE 802 standards.

beneficial in this regard as it enables to control the signals in both frequency and time domains. Thus, the 802.16 standard is based on OFDM, which was selected in preference to competing techniques such as single-carrier (SC) and code division multiple access (CDMA) owing to its superior non line of sight (NLOS) performance. This permits significant equalizer design simplification to support an operation in multipath propagation environments.

The 802.16 PHY provides high flexibility in terms of modulation and coding as SSs may be located at various distances from the BS and hence may experience different signal-to-noise (SNR) ratio. The BS dynamically adjusts the bandwidth, modulation, and coding schemes to overcome the varying SNR and provides improved system performance. OFDM coupled with forward error correction (FEC) techniques, such as Reed–Solomon and convolutional coding, is used when implementing the OFDM PHY. This is the right format to meet the requirements of allocating subcarriers efficiently. In Table 4.1, the rate set along with the modulation types are listed.

IEEE 802.16 also considers optional subchannelization in an uplink. With a subchannelization factor of 1/16, a 12-dB link budget enhancement can be achieved. Sixteen sets of 12 subcarriers each are defined, where one, two, four, eight, or all sets can be assigned to an SS in the uplink. Eight pilot carriers are used when more than one set of subchannels are allocated. To support and handle time variation in the channel, the 802.16 standard provisions optional, more frequent repetition of preambles. In the uplink, short preambles, called mid-ambles, are repeated with a programmable period. In the downlink, a short preamble can be optionally inserted at the beginning of all downlink bursts in addition to the long preamble at the beginning of the frame. A proper implementation of a BS scheduler guarantees minimum required repetition interval for channel estimation.

4.2.2 The MAC

WiMAX offers some flexible features that can potentially be exploited for delivering real-time services. Though the MAC layer of WiMAX has been standardized, there are certain features that can be tuned and made application and channel specific. For example, the MAC layer does not restrict itself to fixed-size frames but allows variable-sized frames to be constructed and transmitted.

The MAC layer of WiMAX comprises three sublayers that interact with each other through the service access points (SAPs) as shown in Figure 4.3. The service-specific convergence sublayer provides the transformation or mapping of external network data, with the help of the SAP. The MAC common

TABLE 4.1

802.16 WiMAX PHY Rate Set

Modulation	Code Rate
QPSK	1/2
QPSK	3/4
16QAM	1/2
16QAM	3/4
64QAM	2/3
64QAM	3/4

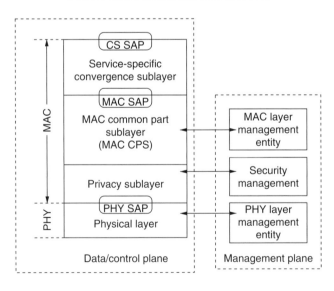

FIGURE 4.3
WiMAX MAC layer with SAPs.

part sublayer receives this information in the form of MAC service data units (MSDUs), which are packed into the payload fields to form MAC protocol data units (MPDUs). Privacy sublayer provides authentication, secure key exchange, and encryption on the MPDUs and passes them over to the PHY layer. Of the three sublayers, the common part sublayer is the core functional layer that provides bandwidth, and establishes and maintains connection. Moreover, as the WiMAX MAC provides a connection-oriented service to the SSs, the common part sublayer also provides a connection identifier (CID) to identify which connection the MPDU is servicing. Let us discuss the different kinds of MAC frame format that WiMAX uses for transmission.

4.2.2.1 MAC Frame Format

In Figure 4.4, the generic MAC frame formats supported by WiMAX (802.16 and 802.16a) for both transport and management information are shown. A generic MPDU consists of a generic MAC frame header (GMH), optional subheaders, payload, and optional forward error correction codes (FEC).

The 6 byte GMH contains details of the entire MPDU as shown in Figure 4.5. The header type (HT) bit at the beginning, when set to 0, indicates that the header is a GMH. The encryption control (EC) bit indicates whether or not the

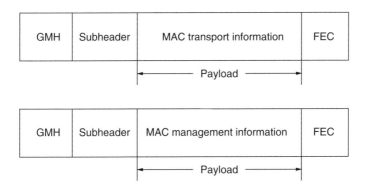

FIGURE 4.4
Generic MAC frame formats.

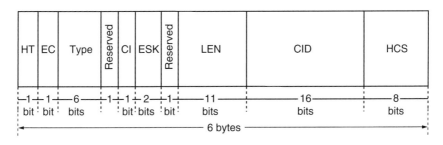

FIGURE 4.5
Generic MAC header.

payload is encrypted and if so, the encryption key sequence (EKS) bits indicate which key was used to encrypt the frame payload. "Type" field reflects the content of payload in terms of whether aggregation, fragmentation, automatic repeat request (ARQ), or mesh feature of the MAC is used. CRC indicator (CI) bit, when set, reveals the presence of error-correction codes at the end. The "LEN" field indicates the number of bytes in the MPDU including the header and the cyclic redundancy check (CRC). The CID defines the connection that the packet is servicing. HCS is appended at the end of GMH, which works as the cyclic redundancy code for the GMH. The optional subheaders are used to define the bits necessary for aggregation, fragmentation, ARQ, and mesh features of the MAC.

The payload inside an MPDU can contain a single MSDU, fragments of MSDUs, aggregates of MSDUs, and aggregates of fragments of MSDUs depending on the MAC rules on aggregation or fragmentation.

4.2.2.2 Aggregation

The common part sublayer is capable of packing more than one complete or partial MSDUs into one MPDU. In Figure 4.6, we show how the payload of the MPDU can accommodate more than two complete MSDUs but not three. Therefore, a part of the third MSDU is packed with the previous two MSDUs to fill the remaining payload field preventing wastage of resources. The payload size is determined by on-air timing slots and feedback received from SS.

To indicate that aggregation is used in the payload of an MPDU, a bit in the "type" field in the GMH is set and the subheaders are used accordingly.

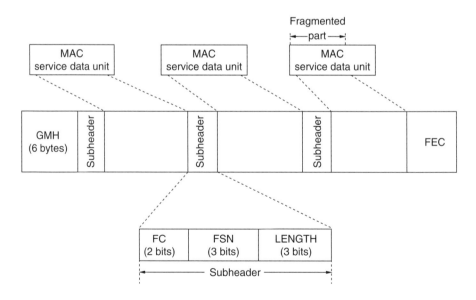

FIGURE 4.6
Multiple MSDUs form an MPDU.

As shown in Figure 4.6, an MPDU can contain multiple subheaders, each followed by either an MSDU or a fragment of an MSDU. FC (fragment control) bits in the subheader are set to 00, if the MSDU is not fragmented and inserted as it is. Otherwise, if the MSDU is fragmented and inserted, FC bits are set to 10, 01, or 11. FSN indicates the fragment sequence number in case an MSDU is fragmented and length field indicates the start of the next subheader in the payload.

4.2.2.3 Fragmentation

The common part sublayer can also fragment an MSDU into multiple MPDUs. In Figure 4.7, we show how a portion of a single MSDU occupies the entire payload of an MPDU. Here, the payload of the MAC packet data unit to be transmitted is too small to accommodate a complete MSDU. In that case, a single MSDU is fragmented and packed into the payload field of the MPDU.

In the case of fragmentation, FC bits are set to 10, if it is the first fragment of the MSDU; 01, if it is the last fragment of the MSDU; and 11, if it is anywhere between first and last fragment of the MSDU. FSN indicates the fragment sequence number as before and the last bit is reserved.

4.2.2.4 Connection Setup and Transmission

A WiMAX BS provides services, including VoIP calls, to the SSs in the cell. The BS can handle multiple VoIP calls simultaneously. Effectively, the last hop of the VoIP path is the WiMAX link that provides the wireless coverage. The identity of each call is maintained by the CID provided by the common part sublayer. As a result, VoIP packets (which are inherently very small) do not have to deal with contention overhead, which greatly increases the efficiency, that is, number of VoIP streams. The connection setup between SSs and BS at the beginning of VoIP packet transmission follows the following three phases.

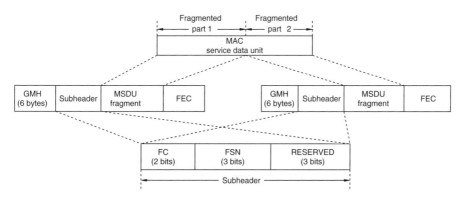

FIGURE 4.7
Single MSDU forms multiple MPDUs.

Phase 1: *Subscriber station requests connection request*

SS that wants a VoIP service stream from the BS transmits the ranging request (RNG-REQ) packet that enables the BS to identify the initial ranging, timing, and power parameters. Service flow parameters requests (bandwidth, frequency, peak, or average rate) are sent next and variable length MSDU indicators are turned on.

Phase 2: *Base station confirms connection*

After receiving connection request from an SS, the BS transmits a ranging response, which provides the initial ranging, timing, and power adjustment information to the SS. VoIP service flow parameters are agreed on and a basic CID is provided to the SS.

Phase 3: *Base station starts transmission of MPDUs*

MSDUs obtained from the MAC convergence sublayer are converted to MPDUs. As needed, MSDUs can be either packed or fragmented to form the desired sized-MPDUs. Since no feedback is received at the start of transmission, the payload and code size agreed at the time of connection is maintained. When a feedback is received, the next awaiting MPDU is formed depending on the type of feedback received. On the reception of the feedbacks, the payload and code sizes are changed. It can be noted that the increase or decrease in payload and code will depend on the ratio of the payload and code.

4.2.2.5 Automatic Repeat Request

ARQ is the means by which a receiver can send a feedback to the transmitter whether an MPDU has been received correctly or not. ARQ mechanism in WiMAX is optional and can be enabled if required. The ARQ frame format along with its subheader is shown in Figure 4.8.

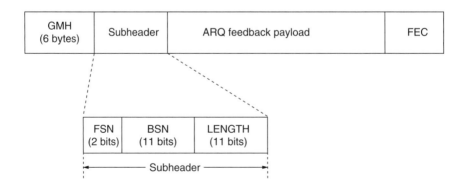

FIGURE 4.8
MAC frame format with ARQ enabled.

To indicate the presence of ARQ feedback payload, a bit in the type field in the GMH is set and the subheader is extended. In the extended subheader, 11 bit block sequence number (BSN) is used instead of the popular 3 bit FSN field as used in the connections without ARQ mechanism.

4.2.2.6 Minislots

The common part sublayer controls the on-air timing on basis of consecutive frames that are divided into time slots (known as minislots in 802.16). The size of these frames and the size of the individual slots within these frames can be varied on a frame-by-frame basis. The users in a WiMAX cell are serviced in a TDMA/TDD manner after their connections are set up. One or many minislots are assigned to every user to service their requests. More formally a minislot is defined as a unit of uplink/downlink bandwidth allocation equivalent to n physical symbols, where $n = 2^m$ and m is an integer ranging between 0 and 7. The number of physical symbols within each frame is a function of the symbol rate. The symbol rate is selected to obtain an integral number of physical symbols within each frame. For example, with a 20 Mbaud symbol rate, there are 5000 physical symbols within an 1 ms frame. This allows effective allocation of on-air resources that can be applied to the MPDUs to be transmitted. Depending on the feedback received from the receiver and on-air PHY layer slots, the size of the MPDU can be optimized. In addition, the division point between uplink and downlink can also vary per frame, allowing asymmetric allocation of on-air time between uplink and downlink if required.

4.3 VoIP

Let us consider the working of a typical VoIP system. A simplified VoIP architecture is shown in Figure 4.9. First, the voice signal is sampled and digitized. Then it is encoded into VoIP frames. There are many popular encoders available, for example, G.711, G.723.1, G.729. The VoIP frames are then packetized and transmitted using RTP/UDP/IP. At the receiver side, the VoIP frames are de-packetized and processed through a playout buffer. The function of the playout buffer is nothing but to smooth out the playing delay, that is, to smoothen the delay jitter caused during transmission through the network. At the end, the voice signal is retrieved from the VoIP frames and is played out at the user's speakers.

As VoIP packets travel through the network, there are evidently some congestion and channel-related losses. Also, the packets suffer delay depending on the congestion at the intermediate routers. Both loss and delay of packets adversely affect the quality of VoIP calls, which is generally expressed in terms of *R*-score.

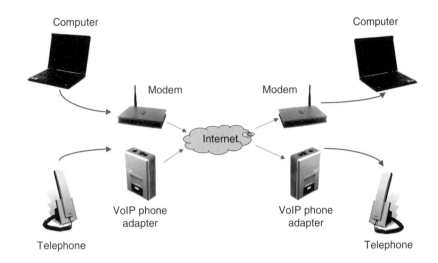

FIGURE 4.9
A simplified VoIP architecture.

4.3.1 Quality of VoIP and *R*-Score

The quality of the reconstructed voice signal is subjective and therefore is measured by the mean opinion score (MOS). MOS is a subjective quality score that ranges from 1 (worst) to 5 (best) and is obtained by conducting subjective surveys. Though these methods provide a good assessment technique, they fail to provide an on-line assessment that might be used for adaptation purpose. The ITU-T E-Model [6] has provided a parametric estimation for this purpose. It defines an *R*-score [6,7] that combines different aspects of voice quality impairment. It is given by

$$R = 100 - I_s - I_e - I_d + A \qquad (4.1)$$

where I_s is the SNR impairments associated with typical switched circuit networks paths, I_e is an equipment impairment factor associated with the losses due to the codecs and network, I_d represents the impairment caused by the mouth-to-ear delay, and A compensates for the above impairments under various user conditions and is known as the expectation factor.

It is to be noted that the contributions to the *R*-score owing to delay and loss impairments are separable. This does not mean that the delay and loss impairments are totally uncorrelated, but their influence can be measured in an isolated manner. Expectation factor covers intangible and almost impossible to measure quantities such as expectation of qualities. However, no such agreement on measurement of expectation on qualities has yet been made and for this reason expectation factor is always ignored while calculating *R*-score. The *R*-score ranges from 0 to 100 and a score of more than 70 usually means a VoIP call of decent quality. The R-score is related to MOS through the following nonlinear mapping [6]:

$$\text{MOS} = 1 + 0.035R + 7 \times 10^{-6}R(R - 60)(100 - R) \tag{4.2}$$

for $0 \leq R \leq 100$. If $R < 0$, MOS takes the value of 1, and similarly, if $R > 100$, MOS takes the value of 4.5.

Among all the factors in Equation 4.1, I_d and I_e are typically considered variables in VoIP [7]. Using default values for all other factors [6], the expression for R-score given by Equation 4.1 can be reduced to

$$R = 94.2 - I_e - I_d \tag{4.3}$$

Let us discuss how end-to-end delay (consisting of codec delay, network delay, and playout delay) and total loss probability (consisting of loss in the network and playout loss at the receiver's decoder buffer) affect the VoIP call quality, that is, the R-score.

4.3.2 Effect of Delay

In a VoIP system, the total mouth-to-ear delay is composed of three components: codec delay (d_{codec}), playout delay ($d_{playout}$), and network delay ($d_{network}$). Codec delay represents the algorithmic and packetization delay associated with the codecs and varies from codec to codec. For example, the G729.a codec introduces a delay of 25 ms. Playout delay is the delay associated with the receiver side buffer required to smoothen the delay for the arriving packet streams. Network delay is the one-way transit delay across the IP transport network from one gateway to another. Thus, the total delay is

$$d = d_{codec} + d_{playout} + d_{network} \tag{4.4}$$

The impact of I_d on voice quality depends on a critical time value of 177.3 ms, which is the total delay budget (one-way mouth-to-ear delay) for VoIP streams. The effect of this delay is modeled as [7]

$$I_d = 0.024d + 0.11(d - 177.3)\mathbf{H}(d - 177.3) \tag{4.5}$$

where $\mathbf{H}(x)$ is an indicator function: $\mathbf{H}(x) = 0$ if $x < 0$, and 1 otherwise.

4.3.3 Effect of Loss

VoIP call quality is also dependent on the loss impairment. Recall, I_e represents the effect of packet loss rate. I_e accounts for impairments caused by both network and receiver's playout losses. Different codecs with their unique encoding/decoding algorithms and packet loss concealment techniques yield different values for I_e. One way to model I_e is to relate I_e to the overall packet loss rate as [7,8]

$$I_e = \gamma_1 + \gamma_2 \ln(1 + \gamma_3 e) \tag{4.6}$$

where γ_1 is a constant that determines voice quality impairment caused by encoding; and γ_2 and γ_3 describe the impact of loss on perceived voice quality for a given codec. Note that e includes both network losses and playout buffer losses, and is modeled as

$$e = e_{network} + (1 - e_{network})e_{playout} \tag{4.7}$$

where $e_{network}$ is the loss probability due to the loss in the network and $e_{playout}$ is loss probability due to the playout loss at the receiver side.

4.3.4 Effect of Delay and Loss on *R*-Score

We show the effect of both delay and loss on *R*-score in Figure 4.10. We show the *R*-scores between 30 and 90 with the jitter buffer being 60 ms. We observe that when the network and jitter loss rate is very low, *R*-score is high, which means that the call quality is also very high. However, with a slight increase in loss, *R*-score decreases rapidly. Moreover, it is observed that with the increase

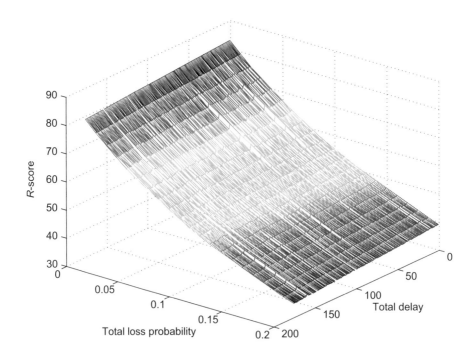

FIGURE 4.10
Sensitivity of *R*-score due to delay and loss.

in network delay, R-score degrades, but at a slower rate. Thus, it is evident that VoIP call quality is more sensitive toward loss than delay. For constant loss and with increase in delay, call quality degrades very slowly. However, for the other case, that is, for constant delay, with increase in loss, the R-score drops significantly.

This difference in sensitivity motivates us to manipulate the loss and delay. Next, we propose an adaptive mechanism that is implemented at the MAC layer of WiMAX. Our objective is to recover as many dropped packets as possible to minimize the loss probability at the cost of increased delay—since loss is more crucial than delay.

4.4 Supporting VoIP over WiMAX

With the sensitivity of VoIP with respect to loss and delay known, let us consider the adaptive schemes at the MAC layer to dynamically construct the MPDUs. Once a connection is set up, the aim behind forming variable-sized MPDUs is such that it strikes a balance between the lost packets and the delay incurred. The final aim is to improve the quality of VoIP calls and at the same time increase the number of streams that can be accommodated.

4.4.1 Packet Restore Probability

When a receiver gets a corrupted packet, it is in no position to correct the errors. However, if some redundant bits in the form of FEC are applied before transmission, then there is a probability that the receiver would be able to detect and possibly correct the errors. The correction capability of these codes depends on the kind and the length of the code used. Let us discuss with respect to the simplest of codes—block codes. In block codes, M redundancy bits are added to the N information bearing bits. (Note that these extra bits are generated using a generator matrix operating on the bits.) If we consider such an MPDU, the resulting bit loss probability is given by [9]

$$b = \sum_{i=M+1}^{M+N} \binom{M+N}{i} b_p^i (1 - b_p)^{M+N-i} \frac{i}{M+N} \qquad (4.8)$$

where b_p is the bit loss probability before decoding and b the decoded bit error probability (BER). The restore probability of such an MPDU with payload size N bits and code M bits is given by $p = (1 - b)^{(M+N)}$. We show three ways to manipulate the packet restore probability.

4.4.1.1 *Decreasing Payload Keeping Code Length Fixed*

Let b be the resulting bit loss probability after decoding of an MPDU with payload of N bits and code length of M bits. If the payload size is decreased to

N' ($N' < N$) keeping the code length fixed, then the resulting bit loss probability after decoding is given by

$$b' = \sum_{i=M+1}^{M+N'} \binom{M+N'}{i} b_p^i (1-b_p)^{M+N'-i} \frac{i}{M+N'} \tag{4.9}$$

Thus, a decrease in payload with the code length fixed lowers the bit loss probability, that is, $b' < b$. If p' is the new packet restore probability then p' is given by

$$p' = (1-b')^{(M+N')} \tag{4.10}$$

As b' and b are close to 0, $(1-b)$ and $(1-b')$ are close to 1. Without loss of generality, it can be said that, for $N' < N$, $p' > p$, that is, with a decrease in payload, packet restore probability increases.

4.4.1.2 Increasing Code Length Keeping Payload Fixed

Similarly, if the code length is increased keeping the payload fixed, the resulting bit loss probability decreases and packet restore probability of MPDUs increases.

4.4.1.3 Increasing Both Payload and Code

The third scheme would be to increase both the payload and the code length. As we know, increasing payload only will increase the resulting BER, so the code length should also need to be increased to compensate for the increased payload.

4.4.2 Enabling ARQ Mechanism

Though the application of FEC enhances packet restore probability, the performance can still be further improved if the optional ARQ mechanism is enabled. The ARQ mechanism at the MAC common part sublayer is enabled by the exchange of control messages between the transmitter and the receiver at the time of connection setup. The ARQ allows feedback to be received at the transmitter side to understand the ongoing call quality and the channel status. By enabling the ARQ mechanism every SS can be made to send a feedback in terms of the packet restore probability from which MAC common part sublayer gets the information whether a packet has been received successfully or not. In addition, these feedbacks give an estimate about the channel status.

Through fast feedback at the MAC layer and use of small packets, the overhead is reduced. The parameters used in the feedback packets are CID, ARQ status (enabled or disabled), maximum retransmission limit, packet restore probability, and a sequence number. The sequence number is used to correlate packets with its response from the BS. If the packet is not received correctly, that is, the packet restore probability is below a certain threshold, then retransmission mechanism is applied. The maximum number of allowed retransmission for a packet is obtained from its corresponding feedback packet. The main advantage of using the retransmission scheme is to lower the loss impairment at the expense of increased delay. For MAC layer retransmissions, a buffer is maintained for every stream at the transmitting WiMAX BS. This buffer helps in temporarily storing the packets unless and until the packets are restored correctly by the receiver. This of course introduces a delay which we denote by d_{queue}. Thus, the total one-way mouth-to-ear delay, as previously given by Equation 4.4, is modified as

$$d = d_{codec} + d_{playout} + d_{network} + d_{queue} \qquad (4.11)$$

To counter this increase in delay, aggregation is used.

4.4.3 Optimal MPDU Size

Since packets often get lost or corrupted during transmission in error-prone wireless channels, ARQ mechanism is usually used to identify and possibly recover the missing frames. In our case, ARQ will play a crucial role in estimating the channel condition and the fate of the MPDUs that have been transmitted. As a result, the round trip time (RTT) becomes crucial in determining the size of the MPDUs. We define RTT as the time difference between the time the last bit of an MPDU is transmitted and the time the acknowledgment for that MPDU is received. Moreover, we assume zero time interval between the transmissions of two consecutive MPDUs, that is, the last bit of an MPDU and the first bit of the next MPDU are transmitted back to back.

Let us now show, how the RTT affects the size of the MPDUs. If we assume that t is the time taken to transmit the MPDU and T is the RTT, then the number of MPDUs already transmitted before the acknowledgment of the first MPDU is received is given by $[T/t]$. It can be noted that t depends on the size of the MPDU and thus there is a trade-off between the goodput (information bits/total bits transmitted) and the delay. If an MPDU is large, the transmission time is large but the overhead due to headers is less, which helps in maintaining a high goodput. If an MPDU is dropped or corrupted owing to bad channel condition, the ARQ mechanism will invoke the retransmission of the large MPDU, which will increase the delay in the transmission. Moreover, by the time the MAC common part sublayer receives the feedback, that is, learns about the channel condition, the transmission of the next MPDU would have already started. If the bad channel condition persists, the probability of the subsequent frame being dropped or corrupted is high. Thus, there will be more retransmissions of large MPDUs under bad channel condition, resulting in severe degradation of goodput compromising the QoS. On the contrary, if the MPDU size is small, the transmission time will be less but the main disadvantage of having small MPDUs is the low goodput due to low payload/overhead ratio. Thus, both large and small MPDUs have their advantages and disadvantages. The advantages of both can be combined by dynamically changing the MPDU size in response to the channel conditions and allocating minislots to obtain the desired level of performance. Next, we discuss how the allocation of the minislots is done on the basis of feedback.

4.4.4 Dynamic Allocation of Minislots

For a G.729a codec, a typical VoIP packet of 60 bytes (40 bytes RTP/UDP/IP header and 20 bytes payload) is fed to the WiMAX MAC layer. At the MAC layer, a minimum overhead is introduced (GMH of 6 bytes for data MPDUs) and FEC codes (depending on the number of retransmissions and codec efficiency) are appended for error recovery. Thus, transmission of an MPDU (consisting of a single MSDU) takes about 8–10 μs. On the contrary, minimum and maximum minislot durations are 1 physical symbol (0.2 μs) and 128 physical symbols (≈26 μs), respectively, with 20 Mbaud symbol rate. Thus, we see that the duration of minislot allocated plays a vital role for VoIP packets. If a single minislot of duration less than the minimum MPDU size is allocated to a session, then there is no way that the MPDU can be accommodated in that minislot. Hence this kind of single-slot allocation cannot be put to effective use. The better option is to allocate multiple minislots to a single VoIP stream to avoid wastage of minislots. Now the question arises how many minislots should be assigned to a single VoIP stream and which scheduling policy should be used to reduce the delay impairment. As each VoIP stream has a delay budget (177.3 ms), the scheduling policy must consider the delay that a VoIP stream has already suffered. Therefore, an appropriate scheduling policy is one in which the BS looks at its buffers for respective streams and calculates the delay by the leading MPDUs in each stream, and assigns the minislots to the stream, which has the highest delay MPDUs. The number of minislots assigned are such that the duration of all the combined minislots are greater than or equal to the MPDU(s) being transmitted.

4.5 Simulation Model and Results

We conducted simulation experiments to evaluate the improvements achieved when the MAC layer features of WiMAX are put to use. Evaluations for adaptive and nonadaptive schemes were done under the same channel conditions for a fair comparison. We assumed a three-state Markov model for the channel. Three states were used to have more granularities in the channel conditions. Each state was characterized by a certain BER: the good state had a BER of 0.01, the medium state had a BER of 0.07, and the bad state had a BER of 1.0. By setting appropriate transition probabilities among these three states, we were able to model different channel conditions for our simulation.

4.5.1 Simulation Parameters

We assume the VoIP streams were generated by a G.729a codec. The other simulation parameters are shown in Table 4.2.

4.5.2 Simulation Results

Here, we present some of the important results. More detailed results can be found in Ref. 2. We assume that the channel remains in the good, medium, and bad state for 30%, 50%, and 10% of the time, respectively. In the adaptive scheme, we used both the ARQ and aggregation schemes, whereas in the nonadaptive scheme, we disabled the ARQ mechanism.

In Figure 4.11a, we present the R-score for both adaptive and nonadaptive schemes. It is seen that with the adaptive scheme, there is an improvement of about 40% in R-score, which indicates that the call quality can be increased in

TABLE 4.2

Simulation Parameters

Simulation Parameters	Values
d_{codec}	25 ms
$d_{playout}$	60 ms
$d_{network}$	70 ms
$e_{playout}$	0.005
WiMAX minislot	2^m PHY symbols
m	0–7
1 ms WiMAX frame	5000 PHY symbols
Symbol rate	20 Mbaud
WiMAX bandwidth	100 Mbps

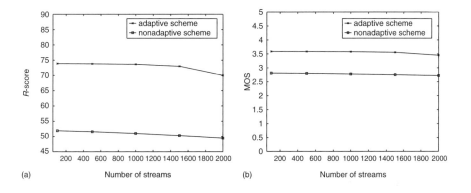

(a) Number of streams   (b) Number of streams

FIGURE 4.11
(a) R-score with adaptive and nonadaptive schemes; (b) MOS with adaptive and nonadaptive schemes.

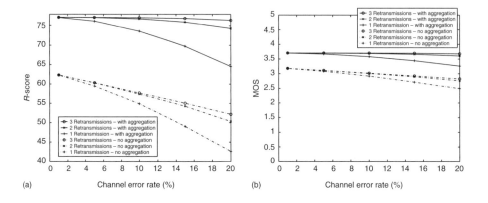

FIGURE 4.12
(a) *R*-score versus error rate with and without aggregation; (b) MOS versus error rate with and without aggregation.

WiMAX using aggregation and ARQ on top of MAC common part sublayer. It can also be noted that with 2000 streams, *R*-score is still above 70, while with 1500 streams it is above 73. In Figure 4.11b, we present the MOS for the same adaptive and nonadaptive schemes. It is observed that with the adaptive scheme, MOS increases significantly (above 3.5) indicating the improvement of call quality of VoIP streams.

Next, we varied the maximum number of retransmissions from 1 to 3. It can be noted that even after the allowed number of retransmissions, the packet might not be restored. In that case, the packet is dropped. With such retransmission schemes, we present the variation of *R*-score and MOS versus channel error rate both with and without aggregation in Figures 4.12a and 4.12b. We fixed the number of VoIP streams at 1000 and gradually increased the channel error rate of the bad state. Retransmissions with aggregation and retransmissions without aggregation are studied separately. It is evident from Figures 4.12a and 4.12b that there is an improvement in *R*-score and MOS specially when the allowed number of retransmissions is 2 and 3. It is observed that with the increase in channel error rates, the rate of decrease is much less for the two or three retransmissions than just one retransmission. It is also evident that the retransmission with aggregation scheme gives better *R*-score and MOS values than the retransmission without aggregation. Thus, it is desirable to bundle both the features (retransmission and aggregation) in WiMAX to improve the call quality in VoIP.

In Figures 4.13a and 4.13b, we show how the *R*-score is affected when the number of VoIP streams is increased. Channel error rate is assumed to be 20%. As expected, retransmission coupled with aggregation yields better *R*-score. Moreover, it is seen that with the retransmission with aggregation, the three-retransmission scheme gives better performance for low and medium load than two-retransmission and one-retransmission schemes, but

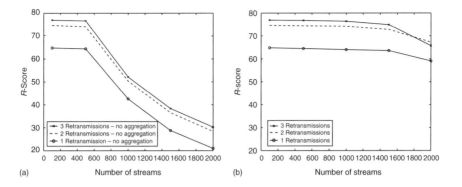

FIGURE 4.13
(a) *R*-score versus the number of streams without aggregation; (b) *R*-score versus the number of streams with aggregation.

the performance degrades with increase in the number of streams. When the load (number of streams) increases, the performance of three-retransmission scheme is worse because with a higher number of retransmissions, a packet suffers increased delay without significant improvement in the packet recovery rate. Allowing at most two-retransmissions at high load is a better choice in this case.

4.6 Conclusions

As new wireless access technologies are being developed, WiMAX is emerging as one of the promising broadband technologies that can support a variety of real-time services. Since extension of VoIP calls over wireless networks is inevitable, we study the feasibility of supporting VoIP over WiMAX.

We discuss a combination of techniques that can be adopted not only to enhance the performance of VoIP but also to support more number of VoIP calls. The proposed schemes, adhering to the MAC layer specification of WiMAX, make use of the flexible features—mainly the size of the protocol data units. We enable the ARQ, use FEC, construct MPDUs by aggregating multiple MSDUs, and dynamically allocate one or multiple minislots to every VoIP call. The performance of the VoIP calls are studied with respect to *R*-score. We exploit the difference in sensitivity of *R*-score toward loss and delay for recovering as many packets as possible at the cost of increased delay. Exhaustive simulation experiments reveal that the feedback-based technique coupled with retransmissions, aggregation, and variable size MPDUs not only increases the *R*-score (and consequently the MOS) but also the number of VoIP streams.

References

1. M. C. Hui and H. S. Matthews, Comparative analysis of traditional telephone and voice-over-Internet protocol (VoIP) systems, *IEEE International Symposium on Electronics and the Environment*, pp. 106–111, May 2004.
2. S. Sengupta, M. Chatterjee, S. Ganguly, and R. Izmailov, Improving R-score of VoIP streams over WiMAX, *IEEE International Conference On Communications (ICC)*, June 2006.
3. http://www.wimaxforum.org, 2006.
4. IEEE Standard for local and metropolitan area networks—Part 16: Air interface for fixed broadband wireless access systems—Amendment 2: MAC modifications and additional physical layer spec. for 2–11 GHz, Std. 802.16a-2003 (Amendment to IEEE Std. 802.16-2001), 2003.
5. S. J. Vaughan-Nichols, Achieving wireless broadband with WiMax, *IEEE Computer*, vol. 37, no. 6, pp. 10–13, June 2004.
6. ITU-T Recommendation G.107, The E-model, a computational model for use in transmission planning, Dec. 1998.
7. R. G. Cole and J. H. Rosenbluth, Voice over IP performance monitoring, *Computer Communication Review*, vol. 31, no. 2, pp. 9–24, 2001.
8. L. Ding and R. A. Goubran, Speech quality prediction in VoIP using the extended E-model, *IEEE GLOBECOM*, pp. 3974–3978, 2003.
9. B. Sklar, *Digital Communications*, 2nd ed. Prentice Hall, NJ, USA.
10. IEEE Std. 802.16-2001 IEEE Standard for local and metropolitan area networks—Part 16: Air interface for fixed broadband wireless access systems, 2002.

5

WiMAX Technology for Home Access

Giselle M. Galván-Tejada and Erickson Trejo-Reyes

CONTENTS

5.1   Introduction

The home access problem has been studied by different authors for several years considering different wireless approaches, including mobile cellular and cordless standards plus satellite and proprietary technologies [1–7]. From these, the development implemented by Nortel at 3.5 GHz [1] perhaps is one of the best known practical examples. However, investment costs have limited the growth of wireless solutions for home access, mainly in highly competitive markets. Nevertheless, the possibility of providing wireless broadband Internet access to residential users has continued as a research topic, leading to the release of IEEE 802.16-2004 standard in 2004 [8], where two spectrum portions (2–11 and 10–66 GHz) were considered for the implementation of fixed broadband communication systems. This standard has received broad acceptation by the several companies that conform the *WiMAX* Forum (for worldwide interoperability microwave access). Thus, in this chapter we discuss issues related to the use of WiMAX technology as an alternative to provide broadband access to residential users.

5.1.1   High Transmission Rates in Wireless Networks

As technology evolves, applications with more bandwidth requirements have also been emerging. Along with these requirements, higher transmission rates are demanded. In a wireless environment, these requirements are not directly and easily fulfilled and even some restrictions can be imposed. The radio channel presents a diversity of propagation conditions, which depend on several factors such as operating frequency, terrain, built-up grade, mobility, coverage, foliage, vegetation, and meteorological phenomena among others. Depending also on applications, the radio channel will present certain characteristics (such as flat or selective behavior to the frequency), which influence the complexity of the technology. Thus, transmission rates are limited by the channel characteristics. Tables 5.1 and 5.2, respectively, summarize the typical transmission rates for some mature cellular mobile and wireless local area networks (WLANs) standards [9].

In wireless residential environments, it is expected that voice, data, and Internet services will demand at least the same quality of service (QoS) as of today's traditional wireline access mechanisms (see Section 5.2.5). Although this requirement complicates the situation for wireless operators, it represents a challenging aspect for the designer of these types of systems.

5.1.2   Coverage

Coverage is another important factor to roll out a wireless network. There are distinct types of environments, for example, indoor, outdoor and outdoor-to-indoor, small business, industrial, residential, academic, large rural environments, etc., all of them implying differences in the desired coverage

TABLE 5.1

Typical Transmission Rates for Cellular Mobile Systems

System	Generation	Era	Transmission Rate
AMPS TACS NTT	1G	Narrowband	2.4 Kbps
IS95 IS136 GSM	2G	Narrowband	64 Kbps
IMT-2000	3G	Wideband	Indoor: 2 Mbps Pedestrian: 384 Kbps Vehicular: 144 Kbps
Broadband wireless	4G	Broadband	1 Gbps

TABLE 5.2

Typical Transmission Rates for Wireless Local Area Networks

Frequency	Transmission Rate	Notes
900 MHz	1–2 Mbps	
18 GHz	6 Mbps	Motorola[a]
2.4 GHz	1.6 Mbps (raw data rates of 11 Mbps)	IEEE 802.11b
2.4 GHz	54 Mbps	IEEE 802.11g <50 m
5 GHz U-NII band	54 Mbps	IEEE 802.11a HIPERLAN/2 <30 m

[a] To date, this product is out of market.

Source: F. Adachi, *The 5th International Symposium on Wireless Personal Multimedia Communications*, vol. 1, pp. 19–26, Oct. 2002. With permission; A. Goldsmit, *Wireless Communications*, Cambridge University Press, 2005. With permission.

(the longer the range, the more complicated it is to achieve high transmission rates—this is one of the reasons to migrate to higher frequencies, but controversially the higher the frequency, the greater the attenuation). Particularly, we are interested in environments for home access applications, so, as it will be addressed in Section 5.4, this chapter will be limited to outdoor applications with different coverage ranges.

5.1.3 Wi-Fi Technology

The migration from traditional wireline to wireless technology has also reached the field of local area networks (LANs), as they have been required to be physically reconfigured or deployed in large office environments. WLANs are so popular today that they are actually preferred in small-area environments.

Thus, a set of standards was developed by the IEEE (Institute of Electrical and Electronics Engineers) and ETSI (European Telecommunications Standards Institute), respectively known as 802.11 and high performance radio LAN (HIPERLAN). Particularly, wireless solutions tested and certified with the IEEE 802.11x standards belong to the technology known as *Wireless Fidelity* or *Wi-Fi*. Wireless LAN technologies today have reached a level of maturity so impressive that their use has been extended even for the deployment of hot spots in public premises. However, Wi-Fi is range-limited and is constrained to low-mobility conditions (i.e., it is better for portable use). In addition, it has been optimized to operate in indoor environments, which makes it impractical for residential use.

5.1.4   WiMAX Arrival

While Wi-Fi and other WLAN technologies were being developed, increasing interest also emerged in the design of wireless metropolitan area networks (WMANs). However, the development of these approaches has lasted more than its corresponding WLAN's counterpart due to the larger ranges and stringent requirements of data, video, and multimedia applications. The task of developing access technologies for this new communications environment was again led by the IEEE and ETSI, as they had worked in the development of their 802.16 and high performance radio MAN (HIPERMAN) standards, respectively.* Today, both organizations work together toward a convergence of both standards [11].

IEEE 802.16 and ETSI HIPERMAN have not been planned for a specific type of network (as it happened with Wi-Fi). Instead, they have been considered to provide an additional set of applications, such as backhaul/backbone, to other wireless networks (e.g., Wi-Fi and cellular), local and metropolitan access, mobility, etc.

Thus, after many discussions the WiMAX Forum joined researchers and manufacturers working together on solving several problems identified, in particular, the channel impairments and the fulfillment of high transmission rates as required by modern applications. The biggest achievement in this effort has been the release of IEEE's standard 802.16-2004 in 2004, best known as fixed WiMAX (see Tables 5.3 and 5.4 for practical specifications of both, base station (BS) and subscriber terminals in a practical WiMAX system).

Depending on the modulation technique and the channel bandwidth, fixed WiMAX can achieve data rates between 32 and 130 Mbps [12]. Another key point of WiMAX is its point-to-point (PTP) and point-to-multipoint (PMP) communication capability [13], which brings the possibility of using it for home access scenarios.

This is where we are interested because, as it will be addressed in Section 5.2, the problem of home access has not been solved well enough, and WiMAX

---

* In the meantime, Wi-Fi has been used to cover the gap of wireless technology between hot spots [9].

TABLE 5.3

Specifications of a WiMAX Base Station

System Capability	Nonline-of-sight operation, PMP deployment
RF Band	3.4–3.6 GHz, 3.6–3.8 GHz (FWA bands)
Center Frequency Resolution	250 kHz
Frequency Stability	+/−4 ppm
Channel Size	3.5, 7 MHz
RF Dynamic Range	>45 dB
Spectral Efficiency	Up to 5 bps/Hz (over-the-air)
Over the Air Rate	Up to 35 Mbps (7 MHz channel)
Latency	6–18 msec
	(depends on channel size, OFDM frame duration)
Maximum Tx Power	+23 dBm
Rx Sensitivity	−96 dBm @ BPSK 1/2
	(10^{-6} BER for 3.5 MHz channel)
IF Cable	Maximum length up to 300 m
	Multiplexed IF, DC power,
	control (Tx/Rx, AGC, APC)
Network Attributes	Transparent bridge, 802.1Q VLAN, 802.1p,
	network prioritization, DHCP, client pass-through
Modulation Coding Rates	Dynamic adaptive modulation (bi-directional)
	Auto-select modulation: BPSK, QPSK, 16QAM,
	64QAM
	Auto-select coding: 1/2, 2/3, 3/4
Over the Air Encryption	DES and AES
MAC	Cell-based PMP deployment
	802.16-2004 compliant PMP
	802.16-2004 packet convergence sublayer mode
	TDMA Access
	Automatic repeat request (ARQ) error correction
Range	Over 3 km/2 miles nonline-of-sight
	Up to 20 km/12.4 miles line-of-sight
Duplex Technique	Dynamic TDD (time division duplex)
	HD-FDD (half duplex frequency division duplex)
Wireless Transmission (PHY)	256 FFT OFDM
Network Connections	TDM (RJ-48c), 10/100 Ethernet (RJ-45)
System Configuration	HTTP (Web) interface, SNMP,
	CLI via Telnet and Local Console
Network Management	SNMP, standard and proprietary MIBs,
	full management by RMS EMS
Power Requirements	Auto-sensing 110/220/240 VAC 50/60 Hz
	Auto-sensing 18–72 VDC, 75 W maximum
Temperature Range	IDU: 0°C–40°C; ODU: −40°C to 60°C
	IDU Short-term: 0°C–55°C for up to 5 h
Wind Loading	Antenna: 137 mph/220 km/h
Physical Configuration	AN-100U Terminal, Radio + Selection of antennas
	All interfaces on front panel
IDU Dimensions	17" × 12" × 1.75"/431.8 mm × 304.8 mm × 44.45 mm

Source: Courtesy of Redline Communications.

TABLE 5.4

Specifications of a WiMAX User Station

System Capability	Nonline-of-sight operation, PMP deployment
RF Band	3.4–3.6 GHz, 3.6–3.8 GHz (FWA bands)
Centre Frequency Resolution	250 kHz
Frequency Stability	+/−4 ppm
Channel Size	3.5, 7 MHz
RF Dynamic Range	>45 dB
Spectral Efficiency	Up to 5 bps/Hz (over-the-air)
Over the Air Rate	Up to 35 Mbps (7 MHz channel)
Latency	6–18 msec (depends on channel size, OFDM frame duration)
Maximum Tx Power	+20 dBm
Rx Sensitivity	−96 dBm @ BPSK 1/2
	(10^{-6} BER for 3.5 MHz channel)
Ethernet Cable	Maximum length up to 75 m
	Multiplexed IF, DC power,
	control (Tx/Rx, AGC, APC)
Network Attributes	Transparent bridge, 802.1Q VLAN, 802.1p,
	network prioritization, DHCP, client pass-through
Modulation Coding Rates	Dynamic adaptive modulation (bi-directional)
	Auto-select modulation: BPSK, QPSK, 16QAM, 64QAM
	Auto-select coding: 1/2, 2/3, 3/4
Over the Air Encryption	DES (traffic) and AES (key exchange)
MAC	Cell-based PMP deployment
	802.16-2004 compliant PMP
	802.16-2004 packet convergence sublayer mode TDMA Access
	Automatic repeat request (ARQ) error correction
Range	Over 3 km/2 miles nonline-of-sight
	Up to 20 km/12.4 miles line-of-sight
Duplex Technique	TDD (time division duplex)
	HD-FDD (half duplex frequency division duplex)
Wireless Transmission (PHY)	256 FFT OFDM
Network Connections	10/100 Ethernet (RJ-45)
System Configuration	HTTP (Web) interface, SNMP, CLI via Telnet
Network Management	SNMP, standard and proprietary MIBs,
	full management by RMS EMS
Power Requirements	PoE injector using 110/220/240 VAC 50/60 Hz
	Meets IEEE 802.3af PoE
Temperature Range	−40°C to 60°C
Wind Loading	Antenna: 137 mph/220 km/h
Physical Configuration	All outdoor units with indoor PoE
	optional IAD assembly
SU-O Dimensions	19 cm × 19 cm × 7 cm

Source: Courtesy of Redline Communications.

is a promising technology in contrast to digital subscriber line (DSL) and broadband cable [14,15]. While being wireless, WiMAX is a straightforward solution in remote areas of difficult access [15,16].

For a wider revision of WiMAX, the reader is referred to several publications in Refs. 13, 17–19. The IEEE 802.16-2004 standard of course is a compulsory reference [8].

5.2 The Home Access Problem

5.2.1 Monopoly of the Local Loop

In a public switched telephone network (PSTN), the local loop is defined as the link between the local exchange (LE) office and the end-user premises. This is illustrated in Figure 5.1.

The local loop (also known as *the last mile*) has traditionally been developed using copper lines and, in numerous cases, has been monopolized by governmental telephone companies that not only hold long-term concession rights but also (in some cases) do not worry for QoS at all.

Despite being the most expensive part of the telephone system (because of one physical link per subscriber and the operating maintenance costs), the local loop is a very attractive solution to be offered due to the number of the potential users [20].

5.2.2 Open Market in the Local Loop

The problems associated with the monopoly of the local loop have motivated a wide discussion on the possibility of opening this segment and creating a

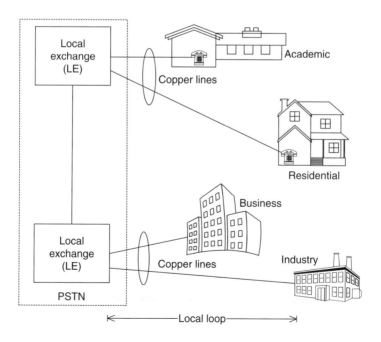

FIGURE 5.1
The local loop into the PSTN.

more competitive scenario where private and public telephone companies are able to offer access to residential users.* The challenges were not only political but also technological. There was the need for a technology that could be easily and quickly implemented and that implies reduced capital expenditure and low operating costs. Thus, in the middle 1990s the research and vendor community presented analysis and alternative solutions to the local loop, for instance, coax cable, power lines, DSL (which is an evolution of conventional wired copper lines), and several variations of radio link technology that are mentioned in Section 5.1. For an excellent summary of all these technologies see Ref. 21.

The revolution around the local loop found an open door in Germany when the country was reunified, because there was a particular interest in quickly and uniformly developing the telecommunications infrastructure along the reunified country. The unique possibility to achieve this goal was by means of wireless technology. Then, the first generation (1G) of systems known as wireless local loop (WLL) or radio in the local loop (RLL) emerged [21,22].

Of all the technologies considered to support the local loop, the WLL approach (nowadays known as fixed wireless access—*FWA* [23,24]) has received wide attention due to the following reasons:

(a) Quick installation

(b) Relatively cheap technology

(c) Relatively easy maintenance

(d) Easy reconfiguration

There are several situations where FWA technology could be an attractive solution to the local loop. They are briefly discussed in the following paragraphs, but the interested reader (especially local access operators) is referred to ETSI's report [22], where they are well documented.

1. Existing operators in a new area. This situation implies delivering local access in the cases of new housing areas near to an existing network or small towns growing and hence requiring basic telecommunications services.

2. Replacement of obsolete copper lines in rural areas. In this case, the main problem is that communications services in rural areas imply high maintenance costs and poor quality and relatively poor revenues. For this reason, it is convenient to group those areas that are relatively near to each other in a radio cluster which will have a central station capable of attending these areas. It is worth noting that the demand of services throughout the several grouped areas could be nonhomogeneous, as some customers will

---

* It is important to note that in addition to residential users, there are business, academic, industrial, and other customers, in principle fixed, that also need access to the PSTN.

require basic voice services, fax and data transmission, better voice quality, etc.

3. Increase capacity of an existing network. This is the case where an existing wireline network has reached its capacity limit and an easy expansion is not possible. If there are customers who require basic voice and narrowband data applications (e.g., fax), they could be changed onto an FWA option leaving their copper lines for other subscribers.

4. New operator in a competitive environment. The entrance of a new player in a competitive environment is a risky and possibly unprofitable move for a potential operator, because the initial costs of rolling out a new network can be so high that in the early stages of operation a few customers will be charged high fees for their services [21]. However, the local loop market is so attractive that operators could still make their business provided FWA technology is carefully implemented. The main challenges that potential operators shall consider are as follows:

   (a) The requirement to serve customers in a variety of scenarios (urban, suburban, and rural areas).

   (b) The need to achieve coverage quickly over the maximum possible area.

   (c) Nonexisting infrastructure (e.g., backhaul) in place.

   (d) The requirements that offered services are technically equivalent to those provided by existing operators, taking into account the possibility that their access solution is wireline.

5.2.3 Structure of an FWA System

The structure of an FWA system is shown in Figure 5.2. Note that an FWA system consists of a number of BSs as in mobile systems and it also exhibits a cellular-like topology, where the BSs are the interface between end users and the PSTN.

You can see that in FWA systems, as opposed to the local loop wireline technologies, there is the possibility of grouping a number of users such that they are served by means of a shared antenna; this reduces not only cost but also installation time.

A particular feature of FWA systems is their fixed nature, which facilitates the use of directional antennas at the CPE (customer premises equipment). Usually, these antennas are installed outdoors at roof level and have typical gains of around 20 dBi [37]. However, there is also the possibility of using indoor antennas, as it will be explained in Section 5.3.3. Figure 5.3 shows two commercial antennas developed for outdoor and indoor FWA scenarios operating with fixed WiMAX technology.

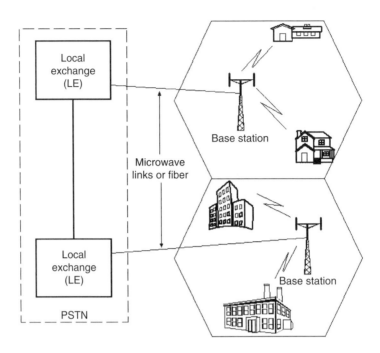

FIGURE 5.2
FWA structure.

FIGURE 5.3
Examples of commercial CPE antennas: Left-hand, outdoor antenna; right-hand, indoor antenna.
(Courtesy of Redline Communications.)

5.2.4   Modern Applications

FWA technology has evolved to offer second-generation (2G) broadband communication, where not only basic voice services are offered, but also data, Internet access, VoIP, video on demand, music, and image downstreaming among others. These applications were anticipated by several authors who stressed the potential of FWA as a serious alternative to compete for the local loop [21,25,26].

2G applications could be exempt for some areas of difficult access (e.g., remote isolated villages), where the fundamental necessity is to provide basic voice communication. Nevertheless, FWA technology could still make the business by providing a wide portfolio of services.

5.2.5   Requirements

Har et al. [27] present a single and illustrative classification of FWA systems from the perspective of bandwidth requirements; in this work two categories are identified:

(a) Narrowband systems as an alternative to basic telephone systems (1G)

(b) Broadband systems that bypass the local loop in order to provide high-speed, interactive services (2G)

Situation (a) corresponds to voice applications and some basic data transmission capabilities. Hence, the required bandwidth is of several kHz. On the contrary, case (b) implies some additional services such as high-speed Internet access and video on demand. Consequently, a few MHz would be needed in situation (b). It is worth noting that 2G FWA systems follow the trend of packet switching (i.e., IP) with guaranteed QoS as it is the case of other broadband access networks.

Specifically, the basic requirements for FWA systems can be outlined as follows [22]:

1. Voice traffic: Typical values of traffic are 70 mErl for residential lines and 150 mErl for business lines.

2. Access network delay: This delay corresponds to that introduced by the radio circuits in the local loop. In spite of the absence of an established maximum value for this delay for FWA, a delay as short as a few tens of milliseconds is recommended to provide an acceptable voice quality.

3. Grade of service (GoS): This figure represents the blocking probability of a system. The value recommended for FWA is 10^{-2}.

4. Lost calls: Under heavy traffic load (even exceeding the designed capacity), established calls should not be lost and blocking in the network should be in accordance with the specified GoS.

5. Service security and authentication: As in any radio system, FWA should consider the implementation of some form of ciphering as a mechanism both to guarantee a secure communication and to authenticate the user into the network.

6. Service transparency: Performance measurements in the communications link should be maintained as in conventional wired networks, for example, a bit error rate, BER $< 10^{-3}$ for voice and a BER $< 10^{-6}$ for data. Other characteristics of standard wired networks, such as number plan, network tones, just to mention a few, should also be transparent to FWA users.

7. Voice, data, and multimedia [25]: FWA systems should be able to provide voice service with wired line quality or better, as well as fax and Internet access at ever high data rates. The growing thrust for using multimedia will also impact the characteristics of FWA systems, and some considerations have been taken into account (see Refs. 26 and 28).

5.3   Propagation Conditions

5.3.1   Line-of-Sight Conditions for FWA

As it is widely described in Ref. 29 and references therein, all propagation environments for terrestrial systems contain different types of terrain and there is no universal criterion that establishes the line that classifies them. Nevertheless, the scientific community has widely adopted the classification of urban, suburban, and rural areas; it is expected that these will be representative of the scenarios where terrestrial wireless systems will serve final home access users (residential, business, academic, industrial, and other fixed users in the local loop).

Now, a line-of-sight (LOS) condition represents the classical situation where the path between transmitter and receiver is clear of obstructions. For this reason LOS conditions are associated with the well-known free-space propagation model whose loss (L_{fs}) is given by expression 5.1:

$$L_{fs} = 32.45 + 20 \log d + 20 \log f \quad \text{(dB)}, \qquad (5.1)$$

where d is the distance (in km) between transmitter and receiver and f the operational frequency in MHz.

Strictly speaking, it is not enough to state that a LOS condition results when there is no obstruction between transmitter and receiver. To have free-space conditions it is necessary that 60% of the first Fresnel zone is clear of

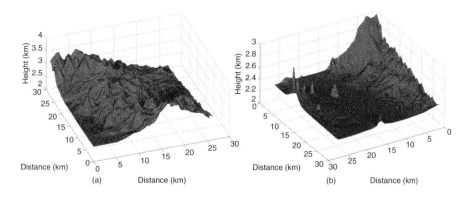

FIGURE 5.4
Example terrain portions (a) irregular terrain (b) regular terrain.

obstacles [29]. At this percentage the diffraction loss caused by one or more obstructions within the first Fresnel zone is equal to the corresponding free space conditions.

In principle, the fixed nature of FWA systems allows the installation of subscriber antennas at poles as high as necessary in order to achieve LOS with the best BS. Nevertheless, the antenna cannot always be raised as high as necessary because it would imply more sophisticated structures to support strong winds. Therefore, in heavily built-up areas or terrains with high variations, LOS conditions cannot always be guaranteed by FWA operators.

To illustrate the aforementioned condition let us consider an experiment on different types of terrain where an FWA system could be in operation. These terrain conditions are representative of an area of 11,655 km² in Mexico (N19W99–N20W100). This area comprises a combination of urban, suburban, and rural zones. The referred area was arbitrarily divided into 16 terrain portions of 728.43 km² each over which the highest point was searched in order to locate a BS (assuming a minimal height of 30 m). SSs were assumed at two possible antenna heights, 8 and 10 m, which could be representative of a typical real scenario. It is also assumed that antennas are directional and properly aligned toward the BS.

In our experiment subscribers with LOS condition were modeled by uniformly distributing 500 users in each terrain portion. The first Fresnel zone was calculated for each user and, depending on the percentage of clearance, the LOS or NLOS (nonline-of-sight) condition was determined for each user.

Figure 5.4 depicts two example terrain portions where the aforementioned calculations were carried out. As it can be seen, terrain in Figure 5.4a is highly irregular, whereas that of Figure 5.4b presents less variations.

Table 5.5 presents the results obtained for the two subscriber antenna heights. As expected, LOS conditions cannot always be guaranteed although in many cases LOS can be achieved. Naturally, the higher the subscriber antenna, the larger the proportion of users in LOS condition.

TABLE 5.5

Number of Users in LOS and NLOS Conditions at 3.5 GHz

	8 m		10 m	
Terrain	Number of Users in LOS Conditions	Number of Users in NLOS Conditions	Number of Users in LOS Conditions	Number of Users in NLOS Conditions
1	252	248	259	241
2	150	350	152	348
3	113	387	116	384
4	192	308	199	301
5	359	141	368	132
6	453	47	458	42
7	160	340	168	332
8	441	59	444	56
9	391	109	396	104
10	344	156	352	148
11	275	225	284	216
12	396	104	400	100
13	285	215	289	211
14	186	314	187	313
15	276	224	286	214
16	324	176	331	169

5.3.2 Fading Phenomenon

The objective of this section is to describe the multipath environment of an FWA system and issues related to the short-term fading. Lee [30] has stated that fixed systems do not exhibit multipath conditions. This significantly simplifies link budget calculations, carrier-to-interference analysis, etc. His assumptions are based on the fact that FWA systems usually employ high antenna gains at user terminals that are physically located in a fixed position. Nevertheless, although directive antennas can reduce the delay spread considerably, provide better conditions for the link budget, etc. [31], the multipath environment does exist as it is depicted in Figure 5.5. Moreover, the FWA channel is time variant, although so slow that this behavior is usually neglected.

The direct and scattering waves arrive at the terminal antenna with different amplitudes, delays, and phases making up a fading pattern. In contrast to cellular mobile systems, the short-term fading characteristics of FWA networks are expected to follow a Rice distribution, since a dominant path could exist [32]. Equation 5.2 shows the Ricean probability density function (pdf) that mathematically models this phenomena:

$$p(x) = \frac{x}{\sigma^2} \exp\left(-\frac{x^2 + s^2}{2\sigma^2}\right) I_0\left(\frac{xs}{\sigma^2}\right); \quad x \geq 0, \qquad (5.2)$$

where s^2 is the so-called noncentrality parameter obtained from the first moments of the component processes, σ^2 the mean power, $x^2/2$ the short-term

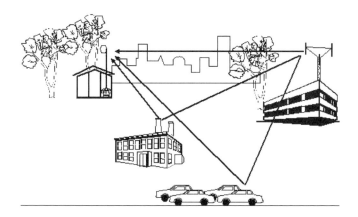

FIGURE 5.5
Multipath environment for FWA.

signal power, and $I_0(\cdot)$ the zero-order modified Bessel function of the first kind [29,32]. By introducing the K-Ricean factor (or simply K-factor, $K = s^2/2\sigma^2$), which represents the difference between the dominant and random paths, the Rice pdf becomes

$$p(x) = \frac{x}{\sigma^2}\exp\left(-\frac{x^2}{2\sigma^2}\right)\exp\left(-K\right)I_0\left(\frac{x}{\sigma}\sqrt{2K}\right).\qquad (5.3)$$

It is worth mentioning that Equation 5.2 (or 5.3) represents the statistics of the signal amplitude fluctuations. However, if the analysis is based on instantaneous power fluctuations, then the pdf that better represents this environment is a Chi-square distribution [33].

There exist numerous countermeasures that overcome the effects of fading: diversity, smart antenna, orthogonal frequency division multiplexing (OFDM), etc. Among these, OFDM has been adopted in the IEEE 802.16-2004 standard, which allows broadband transmission over large ranges (for a wide explication of OFDM see Ref. 34).

5.3.3 Outdoor-to-Indoor Environment

All the scenarios analyzed in Section 5.3.1 assume that subscriber antennas are installed outdoors. But FWA systems also consider the possibility of using indoor residential equipments (see for example Ref. 35). This approach was mainly supported by mobile and cordless operators provided that they could offer an easy solution by selling fixed wireless equipment with an integrated monopole antenna. This solution is very attractive for customers, who can themselves plug in their equipments and any reconfiguration is easily done as well. However, the call rates could be so high (comparable to those of mobile systems) that the approach could become economically unfeasible.

Naturally, this scenario implies NLOS conditions for which some compensation in the antenna gain, receiver sensitivity, or transmission power must be considered.

5.3.4 Propagation Models

Provided that LOS conditions can be achieved for FWA systems, the free space propagation model could directly be applied. Nevertheless, more sophisticated models have been studied in the open literature. For channel modeling simulations, for example, the model known as SUI is considered [36], which includes basic free space loss plus some statistical variations due to possible foliage, vegetation, and obstacles effects. This model has become very popular due to its applicability at 3.5 and 5.8 GHz spectrum bands where new broadband FWA systems based on WiMAX are being planned.

For coverage design, it is necessary to consider physical models [37] where terrain database is used to calculate the loss or received signal strength at each point of the terrain. In this case, the model must include a closed form mathematical expression to deterministically calculate the diffraction attenuation caused by obstructions between transmitter and receiver. Anderson [37] presents a model based on the Epstein–Peterson method [38] for which some other factors are added. In this model, the total loss L_T is simply calculated as:

$$L_T = L_b + C + F + B \quad \text{(dB)}, \tag{5.4}$$

where C represents the loss associated with structures along the path that are not explicitly identified as obstacles, F the loss associated with foliage and B the loss associated with building penetration. The term L_b is the contribution of the loss under free space conditions given by Equation 5.1 and the diffraction loss obtained by the Epstein–Peterson method:

$$L_b = 32.45 + 20 \log d + 20 \log f + \sum_{n=1}^{N} A_n(v, \rho), \tag{5.5}$$

where $A_n(v, \rho)$ corresponds to the diffraction loss calculated for the nth obstacle between transmitter and receiver path. This term is given by Equation 5.6 [29]

$$A_n(v, \rho) = A_n(v, 0) + A_n(0, \rho) + U(v, \rho), \tag{5.6}$$

with

$$A_n(v, 0) = \begin{cases} 0; & v \geq 0.8 \\ 20 \log(0.5 + 6.2v); & 0 \leq v < 0.8 \\ 20 \log[0.5 \exp(0.95v)]; & -1 \leq v < 0 \\ 20 \log\left\{0.4 - \sqrt{0.1184 - (0.1v + 0.38)^2}\right\}; & -2.4 \leq v < -1 \\ 20 \log(-0.225/v); & v < -2.4 \end{cases} \tag{5.7}$$

$$A_n(0, \rho) = 6 + 7.19\rho - 2.02\rho^2 + 3.63\rho^3 - 0.75\rho^4; \quad \rho < 1.4 \qquad (5.8)$$

$$U(v, \rho) = \begin{cases} (43.6 + 23.5v\rho) \log(1 + v\rho) - 6 - 6.7v\rho; & v\rho < 2 \\ 22v\rho - 20 \log(v\rho) - 14.13; & v\rho \geq 2 \end{cases} \qquad (5.9)$$

v is the Fresnel diffraction parameter, which is given by

$$v = -h \sqrt{\frac{2(d_1 + d_2)}{\lambda \, d_1 d_2}}, \qquad (5.10)$$

where d_1 and d_2 are the distances between the obstacle and the transmitter/receiver respectively, λ the wavelength, and h the height taken from the direct path between transmitter and receiver and the edge of the obstacle. Note that all d_1, d_2, λ, and h must be given in the same units. Finally, ρ is a curvature factor expressed by

$$\rho = \left(\frac{\lambda}{\pi}\right)^{1/6} R^{1/3} \left(\frac{d_1 + d_2}{d_1 d_2}\right)^{1/2}, \qquad (5.11)$$

where R is the radius of the curved top of any obstacle. For knife-edge diffraction the terms $A_n(0, \rho)$ and $U(v, \rho)$ become zero and therefore,

$$A_n(v, \rho) = A_n(v, 0). \qquad (5.12)$$

5.4 Feasibility of WiMAX for Home Access

5.4.1 Operation Frequencies for FWA and WiMAX

Initially, different frequencies were considered for FWA systems: 800–900 MHz for systems based on cellular mobile infrastructure [39], 1800 MHz when FWA was based on cordless or DCS-1800 technology [5], and 3.5 GHz for proprietary approaches [1]. Later other frequencies have also been considered such as 5.8 GHz and as high as 20 GHz [40].

In respect of WiMAX, there exist several versions of the standard. The original version, simply known as IEEE 802.16, considered frequencies from 10 to 66 GHz. Then the IEEE 802.16a version lowered the frequency positioning down to the 2–11 GHz range. Finally, the IEEE 802.16d initial profiles specified 2.5, 3.5, and 5.8 GHz spectrum bands.

WiMAX products are mostly considered to operate at frequencies below 10 GHz due to the more favorable propagation conditions, as opposed to the original IEEE 802.16 standard. This trend in WiMAX spectrum positioning also brings the possibility of a smooth migration between the original and proprietary FWA developments onto the more standardized versions of WiMAX.

Hence, from the frequency allocation point of view we can state that WiMAX is suitable for home access.

5.4.2 Evaluation of WiMAX Coverage

5.4.2.1 Cell Radius

In principle, an FWA system is a kind of point to multipoint (PMP) communications solution. Consequently, coverage analyses must take into account the random losses introduced by shadowing. The above is done by including a shadowing term (L_s) in the maximum acceptable loss, L. That is,

$$L = L_{fs} + L_s. \tag{5.13}$$

Note that Equation 5.13 is obtained by assuming shadowing superimposed on free space loss. Then, from Equation 5.13 the cell radius can be obtained as

$$d = 10^{\frac{L - 32.45 - 20\log f - L_s}{20}}. \tag{5.14}$$

Now, shadowing (or slow fading) is mathematically represented by a random variable that is well known to follow a log-normal distribution with standard deviation σ (given in dB), which indicates the local obstruction grade. Then, provided that the probability of slow fading is larger than the average loss (L_{fs}) in at least z dB,

$$P[L_s > z] = Q\left(\frac{z}{\sigma}\right) = Q(t), \tag{5.15}$$

with $Q(\cdot)$ the Q-function, the percentage of desired users at the cell border simply is

$$\% = (1 - Q(t)) \times 100, \tag{5.16}$$

and thus, the value of L_s naturally depends on σ and indirectly on the desired percentage of covered users:

$$L_s = t\sigma. \tag{5.17}$$

Equation 5.14 was used to calculate different cell radii taking, for example, $L = 120$ dB, $\sigma = 4, 6, 8, 10, 12$ dB, and different coverage targets. The results are depicted in Figure 5.6. From this figure it becomes clear how σ impacts on the obtained cell radius for a given percentage of covered area. The larger σ, the shorter the cell range.[*]

According to Ref. 36, the acceptable percentage of FWA system coverage for a given urban, suburban, or rural scenario is between 80% and 90%. If we take this into account (e.g., 90%), and moderate values of $\sigma = 6, 8$ dB then

---

[*] Note that this radius also depends on the maximum acceptable loss, which in turn depends on the equipment capability.

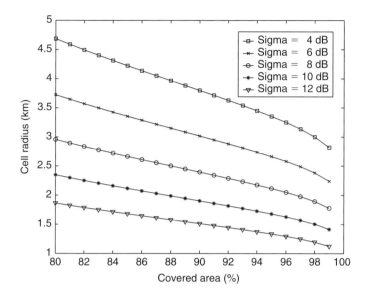

FIGURE 5.6
Cell radius (km).

the cell radius must be between 2.5 and 3.2 km. By comparing these with the results obtained in Section 5.3.1, where the average percentage of users in LOS conditions was less than 60%, it is clear that the terrain dimensions used in Section 5.3.1 must be reduced.

5.4.2.2 Cell Planning

In Ref. 41, a planning study of WiMAX technology in a housing area near Mexico City is presented. The considered application scenario consists of providing wireless access to a residential area located at the bottom of a hill, on top of which there are some high-power TV and radio broadcasting antennas. The above situation imposes a restriction on the planning of the WiMAX BSs location. After a detailed analysis of BS and CPE parameters at 2 GHz an optimal solution was found.

Here, we present an analysis similar to the one presented in Ref. 41, but at a different spectrum band (i.e., 3.5 GHz) and based on the methodology described in Section 5.3.1, where, if we remember, the used terrains provide conditions representative of two types of zones: a quasi-flat terrain and a highly irregular area. In both terrain portions, we located one BS at the highest point. Then, we used Equations 5.4 through 5.12 (Section 5.3.4) for obtaining the referred loss patterns, but assuming knife-edge obstacles and neglecting the terms C, B, and F, provided that only terrain topography information was available. The obtained results are shown in Figure 5.7.

In both cases, the area around to each BS (approximately for a radius of 1 km) is related to a loss level up to 110 dB. As can be seen, the loss distribution

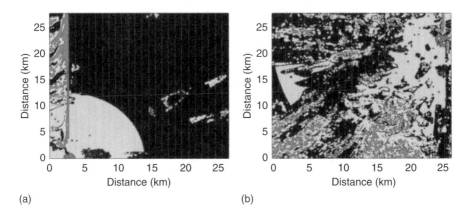

FIGURE 5.7
Loss maps for (a) quasi-flat terrain and (b) highly irregular terrain.

in the case of Figure 5.7a was more uniform than it was for Figure 5.7b, as expected. Consequently, in the quasi-flat terrain an almost well-defined coverage of the BS can be observed (e.g., the semicircle corresponds to a maximum loss of 125 dB), whereas the highly irregular terrain exhibits a more complicated pattern where different loss regions are intermixed (additionally to the semicircle there is up to 155 dB for the black region, etc.).

Finally, if we consider typical values of antenna gains of 19 and 10 dBi for BS and CPE, respectively [37], and a receiver sensitivity of −96 dBm [42], then a maximum loss of 125 dB could be considered as an acceptable target. It becomes clear then that the loss patterns depicted in Figure 5.7 could result in low intensity of the received strength signal and thus more than one BS would be needed in each terrain. After several experiments we realized that for maintaining the link budget at an acceptable level, three BSs would be needed in the quasi-flat terrain portion and four BSs in the highly irregular terrain. The final results are shown in Figure 5.8, showing the final position of the BSs. In this figure, more defined coverage zones can be observed even in the worst situation where the four BSs can attend almost the whole area under study fulfilling the target budget.

5.4.2.3 Hybrid Solution

In Section 5.2.2, referred to the scenario where a group of users (e.g., an apartment building) share some common resources (e.g., antenna, transreceiver, etc.) for the last-mile access and then use another technology to distribute the traffic to their individual premises. A way to address this possibility is a combined WiMAX–Wi-Fi approach [11,14,43], where Wi-Fi is the access mechanism in charge of performing the distribution task. Naturally, the realization of this configuration is limited to the maximum number of users that Wi-Fi can support.

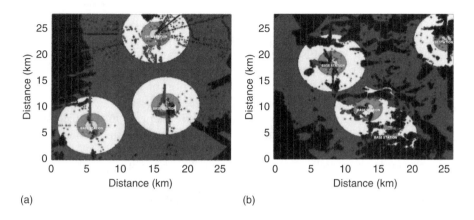

FIGURE 5.8
Final positions of base stations (a) quasi-flat terrain and (b) highly irregular terrain.

5.4.3   Transmission Rates of WiMAX versus the FWA Requirements

As it was addressed in Section 5.2.4, the current applications for FWA systems include more than basic voice services. In general terms, it is expected that modern broadband data rates shall be around 1.5 and 2 Mbps. More specifically, the requirements for different customers can vary. As Webb [21] points out, the European group of ETSI named *Broadband Radio Access Networks* studied these requirements and concluded that the data rates for FWA applications could be: from 1.5 to 6 Mbps for video; 2 Mbps for Web browsing; 10 Mbps for corporate access; and up to 25 Mbps for LAN interconnection.

As far as WiMAX is concerned, this technology provides broadband data rates up to 70 Mbps [19] depending on the range and channel bandwidth, which supports several connections at equivalent DSL rates [11,14]. Therefore, WiMAX can support modern home access application requirements while fulfilling the demand of high data rates.

5.5   Possible Improvements

Some improvements can be implemented by operators interested on providing last-mile wireless access based on WiMAX technology. Many of them have already been addressed in Ref. 13, including spatial multiplexing/precoding, interference cancellation, adaptive sub-carrier/power allocation and hybrid automatic repeat request (ARQ). Among these, let us briefly comment some ideas about spatial multiplexing, which is based on the use of smart antennas. The topic of smart antennas has attracted the attention of the scientific community for different reasons. This technology can reduce the interference levels and improve the quality of the receiving signals. Also, the coverage range of BSs could be extended or equivalently the transmitted power could be reduced. Other sophisticated multiple access techniques such as space division multiple access (SDMA) could be implemented to achieve larger system capacities [44]. In fact, this aspect is also considered in the IEEE 802.16-2004 standard [8] as a possibility. In case that system capacity could be at its limit, another possibility would be the implementation of antenna arrays at the subscriber premises, as demonstrated in Ref. 45.

Another possible improvement is the use of frequencies above 10 GHz due to greater bandwidths. Applications based on these frequencies would be limited to relatively short distances due to the attenuation caused by rain and absorption due to water vapor and oxygen. This type of applications could be conducted in areas of high traffic demand [40].

Finally, other future propagation models could be used. Although there are some studies of the loss caused by foliage, some other measurements campaigns must be conducted first for different frequencies and type of vegetation.

5.6 Conclusions

The use of WiMAX technology for the home access or last-mile has been addressed in this chapter. A review of the home access problem was initially given together with the proposed solutions published for more than one decade by several authors. Typical requirements were also addressed. Wireless technologies were presented as predecessors of WiMAX.

A theoretical framework of LOS conditions was discussed in the context of FWA systems. Simulations were carried out over realistic terrain scenarios to evaluate the percentage of users whose link could be in free space conditions. As expected, many cases presented LOS paths, but some other cases were considerably affected by obstructions.

The feasibility of WiMAX for home access was analyzed from different perspectives, including data rate, frequency, coverage, and cell planning. The results obtained in this evaluation suggest that WiMAX is suitable for the last mile, both in its conventional form of providing direct connection to the end user, or by combining Wi-Fi and WiMAX for some specific scenarios.

Finally, it is important to highlight that other environments could also be studied in the planning stage. This is the case of outdoor-to-indoor propagation provided that today there are numerous subscriber equipment models designed to operate at indoor scenarios. In this case, propagation models extended to this environment should be considered.

Acknowledgments

The authors would like to acknowledge and thank Edgar Ochoa and Carolyn Anderson from Redline Communications for their support and the publishing of confidential material from Redline Communications, a most valuable help on better understanding the contents of this chapter.

References

1. R. McArthur, Ionica fixed radio access system, *IEEE Colloquium on Local Loop Fixed Radio Access*, pp. 5/1–5/11, 1995.
2. F. G. Harrison, Microwave radio in the British telecom access network, *British Telecommunications Engineering*, vol. 8, pp. 100–106, 1989.
3. H. M. Sandler, CT2 radio technology for low power fixed wireless access, *IEEE International Symposium on Personal, Indoor and Mobile Radio Communications*, vol. 3, pp. 1133–1138, 1995.
4. S. Kandiyoor, P. Van de Berg, and S. Blomstergren, DECT: Meeting needs and creating opportunities for public network operators, *IEEE International Conference on Personal Wireless Communications*, pp. 28–32, 1996.
5. M. Lotter and P. Van Rooyen, CDMA and DECT: Alternative wireless local loop technologies for developing countries, *IEEE International Symposium on Personal, Indoor and Mobile Radio Communications*, Helsinki, Finland, pp. 169–173, 1997.
6. R. Westerveld and R. Prasad, Rural communications in India using fixed cellular radio systems, *IEEE Communications Magazine*, vol. 32, no. 10, pp. 70–77, Oct. 1994.
7. D. Anvekar, P. Agrawal, and T. Patel, Fixed cellular rural networks in developing countries: A performance evaluation, *IEEE International Conference on Personal Wireless Communications*, pp. 33–38, 1996.
8. IEEE 802.16-2004, *IEEE Standard for Local and Metropolitan Area Networks – Part 16: Air Interface for Fixed Broadband Wireless Access Systems*, Oct. 2004.
9. F. Adachi, Evolution towards broadband wireless systems, *The 5th International Symposium on Wireless Personal Multimedia Communications*, vol. 1, pp. 19–26, Oct. 2002.
10. A. Goldsmit, *Wireless Communications*, Cambridge University Press, Cambridge, 2005.
11. N. Fourty, T. Val, P. Frase, and J. Mercier, Comparative analysis of new high data rate wireless communication technologies: From Wi-Fi to WiMAX, *2005 Joint International Conference on Autonomic and Autonomus Systems and International Conference on Networking and Services*, pp. 66–71, 23–28 Oct. 2005.
12. K. Wongthavarawat and A. Ganz, IEEE 802.16 based last mile broadband wireless military networks with quality of service support, *IEEE Military Communications Conference*, vol. 2, pp. 779–784, 13–16 Oct. 2003.
13. A. Ghosh, D. R. Wolter, J. G. Andrews, and R. Chen, Broadband wireless access with WiMAX/802.16: Current performance benchmarks and future potential, *IEEE Communications Magazine*, vol. 43, no. 2, pp. 129–136, Feb. 2005.
14. F. Behmann, Impact of wireless (Wi-Fi, WiMAX) on 3G and next generation— An initial assessment, *2005 IEEE International Conference on Electro Information Technology*, pp. 1–6, 22–25 May 2005.
15. S. Jindal, A. Jindal, and N. Gupta, Grouping WI-MAX, 3G and WI-FI for wireless broadband, *The First IEEE and IFIP International Conference in Central Asia on Internet*, pp. 1–5, 26–29 Oct. 2005.

16. N. Chauville, D. Chatelain, and B. J. van Wyk, WiMAX access over GSM/GPRS in rural areas, *IEEE 12th International Symposium on Electron Devices for Microwave and Optoelectronic Applications*, pp. 106–109, 8–9 Nov. 2004.

17. C. Eklund, R. B. Marks, K. L. Stanwood, and S. Wang, IEEE standard 802.16: A technical overview of the wireless MANTM air interface for broadband wireless access, *IEEE Communications Magazine*, vol. 40, no. 6, pp. 98–107, June Berkely, CA, 2002.

18. S. J. Vaughan-Nichols, Achieving wireless broadband with WiMAX, *Computer*, vol. 37, no. 6, pp. 10–13, Berkely, CA, June 2004.

19. D. Sweeney, *WiMAX Operator's Manual: Building 802.16 Wireless Networks*, Apress, Berkeley, CA, June 2004.

20. G. Calhoun, *Wireless Access and the Local Telephone Network*, Artech House, Norwood, MA, 1992.

21. W. Webb, *Introduction to Wireless Local Loop: Broadband and Narrowband Systems*, 2nd ed. Artech House, Boston, London, 2000.

22. ETR 139, Radio Local Loop, ETSI, Nov. 1994.

23. C. Hart, Fixed wireless access: A market and system overview, *Electronics and Communication Engineering Journal*, pp. 213–220, Oct. 1998.

24. S. Nomoto, M. Asano, Y. Ito, M. Nakamura, and T. Abe, Fully adaptive MODEM up to 1024 QAM for 26 GHz broadband fixed wireless access systems with mesh topology, *IEEE 1st International Symposium on Wireless Communication Systems*, pp. 100–104, 20–22 Sept. 2004.

25. J. Haine, Overall requirements for fixed radio access in the local loop, *IEEE Colloquium on Local Loop Fixed Radio Access*, pp. 1/1–1/7, Dec. 1995.

26. J. Haine, HIPERACCESS: An access system for the information age, *Electronics and Communication Engineering Journal*, pp. 229–235, Oct. 1998.

27. D. Har, C. Xu, and H. H. Xia, Propagation modeling for wireless local loop channel, *International Journal of Communications Systems*, vol. 13, no. 3, pp. 231–241, May 2000.

28. ATM radio-in-the-local-loop (RL/A), *Proceedings of the Wireless Broadband Communications Workshop*, Brussels, Belgium, 29, Sept. 1997.

29. D. Parsons, *The Mobile Radio Propagation Channel*, Halsted Press, New York, 1992.

30. W. C. Y. Lee, Spectrum and technology of a wireless local loop system, *IEEE Personal Communications Magazine*, pp. 49–54, Feb. 1998.

31. A. G. Burr, A spatial channel model to evaluate the influence of directional antennas in broadband radio system, *IEEE Colloquium on Broadband Digital Radio: Challenge of the Radio Environment*, pp. 8/1–8/8, 1998.

32. R. Steele and L. Hanzo (Editors), *Mobile Radio Communications*, Prentech Press, London, 1992.

33. J. P. Linnartz, *Narrowband Land-Mobile Radio Networks*, Artech House, Norwood, MA, 1993.

34. A. R. S. Bahai, *Multicarrier Digital Communications: Theory and Applications of OFDM*, 2nd ed. Springer, New York, 2004.

35. M. A. Abumahlula, *DECT Technology for Wireless Local Loop*, PhD Thesis, University of Bradford, 1997.

36. IEEE 802.16.3.c, *Channel Models for Fixed Wireless Applications*, Broadband Wireless Access Working Group, 2001.

37. H. R. Anderson, *Fixed Broadband Wireless System Design*, John Wiley and Sons, West Sussex, England, 2003.

38. J. Epstein and D. W. Peterson, An experimental study of wave propagation at 850 MC, *Proc. IRE*, vol. 41, no. 5, pp. 595–611, 1953.
39. Rec. 757, Basic Requirements and Performance Objectives for Cellular Type Mobile Systems Used as Fixed Systems, CCIR, 1992.
40. K. Nidaira, T. Shirouzu, M. Baba, and K. Inoue, Wireless IP access system for broadband access services, *IEEE International Conference on Communications*, vol. 6, pp. 3434–3438, 20–24 June 2004.
41. J. Garcia-Fragoso and G. M. Galvan-Tejada, Cell planning based on WiMAX standard for home access: A practical case, *IEEE 2nd International Conference on Electrical and Electronics Engineering and XI Conference on Electrical Engineering*, Mexico City, Mexico, pp. 89–92, 7–9 Sept. 2004.
42. www.redlinecommunications.com
43. J. Nie, X. He, Z. Zhou, and C. Zhao, Communication with bandwidth optimization in IEEE 802.16 and IEEE 802.11 hybrid networks, *IEEE International Symposium on Communications and Information Technology*, vol. 1, pp. 26–29, 12–14 Oct. 2005.
44. G. M. Galvan-Tejada and J. G. Gardiner, Theoretical model to determine the blocking probability for SDMA systems, *IEEE Transactions on Vehicular Technology*, vol. 50, no. 5, pp. 1279–1288, Sept. 2001.
45. G. M. Galvan-Tejada and J. G. Gardiner, Use of an antenna array at customer stations to improve the capacity in a WLL network, *International Journal of Communication Systems*, vol. 13, no. 3 (WLL special issue), pp. 243–253, 2000.

6

WiMAX Enables Cyber Extension to Rural Communities

K.R. Santhi, G. Senthil Kumaran, and Albert Butare

CONTENTS

6.1   Introduction

Access to Internet is of prime importance as it has turned into a fully con-
verged network delivering voice, audio, image, and video in addition to
data. WiMAX extends the reach of IP broadband metropolitan fiber networks
well beyond the relatively small local area network (LAN) coverage of Wi-Fi
in offices, homes, or public access hot spots to rural areas. WiMAX is expected
to provide flexible, cost-effective, standards-based means of filling existing
gaps in broadband coverage, and creating new forms of broadband services
not envisioned in a "wired" world. WiMAX is widely viewed as a "great
wireless hope" for outdoor wireless services, and the disruptive potential of
WiMAX looks set to revolutionize the entire broadband wireless access (BWA)
industry. In this part, we analyze the need for a BWA technology and provide
you with the advantages of WiMAX as a last-mile solution.

6.2   Need for a Standard-Based Broadband Wireless Access

BWA has obvious advantages for any place that is not wired already and
it is a wireless alternative for wired broadband access. It is a carrier-based

solution in both rural and congested cities where technical, physical, and economic constraints of wireline services simply fail. The following are the reasons that support the need for a standard-based BWA.

6.2.1   Increasing Demand for Broadband Access

BWA is a solution to meet the escalating demand-pull of both broadband-starved regions and the massive growth of the use of Internet. The emerging BWA will provide realistic services, including rural areas as a last-mile solution; urban deployments as an alternative to low-cost, self-installed fixed broadband (digital subscriber line (DSL) and cable) services; and in developing regions where there is little or no wired infrastructure. Also security concerns have inspired many cities to embrace broadband wireless solutions for surveillance.

6.2.2   Lack of Available Options in Underdeveloped Areas

Although broadband has been available for some time, access for most people is still limited. The cost and complexity associated with traditional wired cable and telephone infrastructure have resulted in significant broadband coverage gaps in many developed and developing countries. Mostly, in any country the residences are situated in the suburban parts that are too far away from phone network central offices to get a usable signal. Rural areas do not have a wired broadband network. Filling these coverage gaps involves have involved a number of proprietary solutions for BWA, which are expensive. But with standard-based BWA these barriers can be lifted.

6.2.3   Problems with Proprietary BWA Solutions

Early attempts to BWA deployments relied on some type of proprietary solution and it has fragmented the market without providing significant economies of scale. In an ideal world, the vendors would love to be able to continue in their higher margin proprietary business. But a standard-based technology will make different vendors products compatible and access much cheaper. Only a standards-based approach will bring BWA to the masses. But the downside to a standards-based approach is that the process of definition, ratification, and product certification is time consuming [1].

6.2.4   Improving Anytime Anywhere Access to Data

For anytime anywhere access, a technology that operates at greater distances and over a greater range of frequencies is needed. BWA sounds great as long as it really does cost less to use than our earthbound skein of wires, fibers, and cables. For this to happen the Wi-Fi technology needed to evolve into more of a carrier technology, deployed by a service provider, and needed to exploit a lot more spectrum options or we adopt a standard-based BWA technology.

6.2.5   Constraints with Wired Broadband

Currently, there are cable and DSL broadband access services in the market-place. But, their practical limitations in features and deployment have prevented them from reaching many potential broadband Internet customers [2]. The wired broadband connection provided by cable and DSL is an all-consuming and expensive process. A large number of areas throughout the world currently are not able to access broadband connectivity. Tradition-ally, DSL can only reach about 18,000 feet (3 miles) from the central office switch, and this limitation means that many urban and suburban locations may not be served by DSL connectivity. The limitation of cable is that many older cable networks have not been equipped to offer a return channel, and converting and deploying these networks to support high-speed broadband can be expensive.

6.2.6   Improving Business Models

Standard-based BWA deployments will not only increase the competition in the fixed broadband and cellular data market, but also have a deep and long-lasting effect on how broadband services are used and marketed on the overall broadband value chain and on the business models of the network operators and service providers. Service providers do not have to rely on a single vendor and so economies of scale are possible.

6.3   Network Architecture of WiMAX

6.3.1   Working Principle

Wireless broadband access is set up like cellular systems, using base sta-tions (BSs) that service a radius of several miles/kilometers. BSs are either directly wired to the Internet or use WiMAX links to other BSs that are so connected. Clients devices/customer premise equipment (CPE) are gener-ally small, building-mounted antenna/transceiver systems. The signal is then routed via building wireless local area networks (WLANs) or standard Eth-ernet cable either directly to a single computer, or to an 802.11 hot spot or a wired Ethernet LAN that connects the BS to a customer [2].

6.3.2   WiMAX Deployment Model

There are several ways WiMAX can be deployed. One is high-bandwidth, LOS, point-to-point (PTP) backhaul, for example, from 2G/3G sites or Wi-Fi hot spots. A WiMAX tower connecting to another WiMAX tower using line-of-sight (LOS), microwave link is referred to as a backhaul. Another is

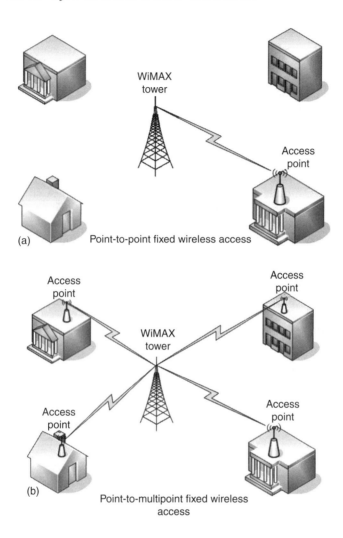

FIGURE 6.1
(a) Point-to-point topology. (b) Point-to-multipoint deployment of WiMAX.

"metro Ethernet," where bandwidths from 10 Mbps and upward are provided, on point-to-multipoint (PMP) basis, with a typical cell radii of up to 5 miles/8 km under nonline-of-sight (NLOS) conditions and provide access to small business units or residential areas. Figure 6.1 shows the fixed wireless access (FWA) topologies of WiMAX.

6.3.3 Cell Structure

Each BS provides wireless coverage over an area called a cell. As with conventional cellular mobile networks, the BS antennas can be omnidirectional, giving a circular cell shape, or directional for PTP use or for increasing the network's capacity by effectively dividing large cells into several smaller sectoral areas. People tend to think that you can put one WiMAX tower on a hillside and beam around the entire city, and that is certainly not the case. When you fill up a cell, you use up the capacity meaning that providers will still have to add towers as demand grows, just as they do in traditional cell phone networks.

6.3.4 WiMAX Base Station (BS)

WiMAX BS is similar in concept to a cell phone tower. A single WiMAX BS/tower can transmit voice, video, and data signals across distances of up to 50 km from a central tower with unobstructed LOS at rates as high as 70 Mbps. A WiMAX tower can connect directly to the Internet using a high-bandwidth, wired connection. A single tower's ability to cover up to 3000 square miles is what allows WiMAX to provide coverage to remote rural areas.

6.3.5 Customer Premise Equipment (CPE)

The first generation CPE or access point (AP) will be outdoor installable subscriber station (SS) akin to a small satellite dish. The second generation CPE will be indoor self-installable modems similar to a cable or DSL modem [3]. The third generation CPE will be a small box or a PCMCIA card, or they could be built into a laptop like Wi-Fi access.

6.3.6 Mesh Topology

A further bonus of WiMAX is that it supports mesh networks. This means that WiMAX-enabled devices can act as relays, passing signals from one device to another until they reach a WiMAX BS from which they can enter the wired Internet. Relaying like this greatly extends the potential range of an AP and allows network to grow in an organic fashion. Figure 6.2 shows the same.

6.4 Advantages of WiMAX

Compared to the previous generation of technologies WiMAX offers several advantages.

6.4.1 Economic Advantages

Companies can realize the following economic benefits directly or indirectly by implementing WiMAX.

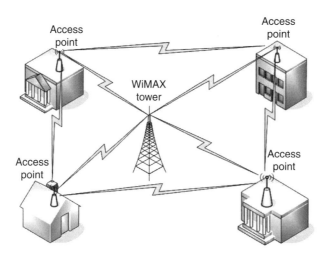

FIGURE 6.2
Mesh configured fixed wireless access.

1. Ease of installation in difficult-to-wire areas: The ability to quickly provision service, even in areas that are hard for wired infrastructure to reach; for example, if rivers, free ways, or other obstacles separate buildings you want to connect, a wireless MAN (WMAN) solution may be much more economical than installing physical cable or leasing communications circuits.

2. Scalability: Imagine hundreds of hot spot users at a 5-day conference trying to access the network. Accessing the LAN would be no problem through 802.11 hot spots [4]. But what if those users want to simultaneously access the Internet? The hotel might have a limited broadband connectivity use; however, for those 5 days, it needs a lot more bandwidth. Instead of waiting weeks for a T1 or DSL line, WiMAX access can be quickly and easily set up at new and temporary sites.

3. Mobile data services: 70 Mbps of shared bandwidth to mobile devices, possibly in unlicensed spectrum, allows a user to roam around on their wireless ISP seamlessly from their home Wi-Fi network. So, data connectivity becomes increasingly limited to the individual subscriber like mobile phones are being used and less dependent on a specific location.

4. Economies of scale: At the moment, BWA vendors rely on proprietary, custom-built chipsets, which have kept equipment costs high and customers away. With standardization comes the ability to mass produce products, so research and development costs decline along with manufacturing expenses. This could pave the way for

lower cost services, which WiMAX proponents say will make BWA feasible for mass deployment.

5. Commercialization: Commercialization and cost reduction of various smart antenna technologies will allow operators to limit requirements for BWA wireless infrastructure, making deployments cheaper [5].

6.4.2   Technical Advantages

WiMAX-based solutions include many other advantages, such as robust security features, good QoS, and mesh and smart antenna technology that will allow better utilization of the spectrum resources.

1. Subchannelization allows scalability: Subchannelization is an optional feature in orthogonal frequency division multiplexing (OFDM) 256, which is generating a lot of interest from operators. It allows an SS to concentrate its transmit power on a subset (subchannel) of the total OFDM subcarriers, leading to link budget improvements in the uplink. These translate into coverage and capacity benefits. Multiple SSs can be scheduled to transmit simultaneously on different subchannels.

2. Adaptive modulation provides throughput: By using a robust modulation scheme, IEEE 802.16 delivers high throughput at long ranges with a high level of spectral efficiency that is also tolerant of signal reflections. Dynamic adaptive modulation allows the BS to trade off throughput for range [6]. For example, an SS close to the BS can use 64QAM modulation, while the weaker signal from a more remote subscriber might only permit the use of 16QAM or QPSK. The medium access control (MAC) layer can use different modulation methods for each subscriber's downlink and uplink burst. Figure 6.3 shows coverage versus modulation.

3. Mesh topology increases coverage: In addition to supporting a robust and dynamic modulation scheme, the mesh topology and smart antenna techniques greatly enhance coverage in extreme environments by using multiple antennas to create "transmit" and "receive diversity" [6]. The major benefit of BWA is the ability to provide coverage for a wide variety of uses.

4. Security: Privacy and encryption features are included in WiMAX to support secure transmission and provide authentication and data encryption [6].

5. Increased reliability: A problem inherent in wired networks is the downtime due to cable faults. Moisture erodes metallic conductors. These imperfect cable splices can cause signal reflections that result in unexplainable errors. The accidental cutting of cables can also

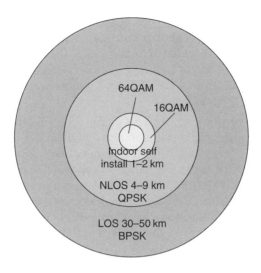

FIGURE 6.3
Relative cell radius versus adaptive modulation.

bring a network down quickly. These problems interfere with the users' ability to utilize network resources causing havoc for network managers. The advantage of WiMAX networking is that it will likely experience fewer problems because less cable is used.

6.5 Applications of WiMAX as a Last-Mile Solution

This section will examine some of the specific application-driven opportunities engendered by WiMAX and its derivative technologies. WiMAX offers an assortment of engaging applications and services for both residential and business end users.

1. Last-mile solutions: Practical limitations prevent cable and DSL technologies from reaching many potential broadband customers. The absence of LOS requirement, high bandwidth, and the inherent flexibility and low cost of WiMAX prove useful in delivering broadband services to rural areas where it is cost-prohibitive to install landline infrastructure.

2. Cellular backhaul: WiMAX can play a role in enabling mobile operators to cost effectively increase backhaul networks for cellular BSs, bypassing the public switched telephone network (PSTN) [2]. Cellular service providers will also look at wireless backhaul as a more cost-effective alternative.

3. Wi-Fi backhaul: One of the obstacles for continued hot spot growth is the availability of high-capacity, cost-effective backhaul solution. WiMAX provides backhaul connections to the Internet for Wi-Fi hot spots. It will also fill the coverage gaps between hot spots, and allow users to connect to a wireless ISP even when they roam outside their home or business office.

4. Broadband on demand: WiMAX enables a service provider to offer instantly configurable on "demand" high-speed connectivity for temporary events including trade shows that generate hundreds or thousands of users for 802.11 hot spots. It makes possible for the service provider to scale-up or scale-down service levels, within seconds of a customer request. "On demand" connectivity also benefits businesses that move their operations frequently, such as a construction company with offices at each building site.

5. Underserved areas: WiMAX is a natural choice for Internet access to poor underserved rural regions. In such areas, local utilities and governments work together with a local wireless internet service provider (WISP) to deliver service.

6. Best-connected wireless service: The IEEE 802.16e introduces nomadic capabilities that will allow users to connect to a WISP even when they roam outside their home or business area, or go to another city that has a WISP. With 802.16e, users will be able to stay connected by 802.11 when they are within a hot spot, and then connected by 802.16 when they leave the hot spot but are within a WiMAX service area.

7. Public safety services: Support for nomadic services and ability to provide ubiquitous coverage over large area provides a tool for surveillance which is just one potential application for broadband wireless. For example, police departments want high-quality photos to be distributed to their vehicles in real time [7]. Officers want to send the photos back to headquarters for immediate analysis, to analyze potential suspects. Figure 6.4 shows the various applications of WiMAX.

6.6 WiMAX Design Challenges

802.16 standards have defined too many design options for WiMAX and industry leaders face challenges to select one from the many available.

6.6.1 TDD, FDD, or H-FDD

As WiMAX supports TDD, FDD, and H-FDD, the design of the system also varies depending on which scheme is selected. If the system will be FDD,

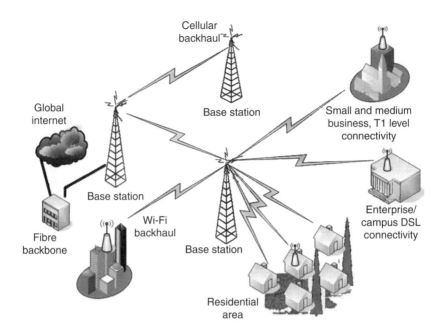

FIGURE 6.4
Various applications of WiMAX.

two complete radios (including synthesizers) operating simultaneously on different frequencies will be required. The cost is high and there are external filtering requirements.

If H-FDD is selected, it has a single radio (and single synthesizer) and similar costs to TDD. But the key concern is that the synthesizer must be able to switch between the transmitter and receiver within 100 ms, which is a constraint [8].

Many industry leaders expect that BSs will use full FDD mode due to its higher throughput, while the SSs will use lower cost H-FDD or TDD [8].

6.6.2 Error Vector Magnitude

Error vector magnitude (EVM) must be 6 dB higher for 802.16 (−31 dB) than for 802.11 (−25 dB). This has a number of implications. First, WiMAX system typically transmits at a higher power than an 802.11 system. Hence, the power amplifiers (PAs) must deliver more power, they must be more linear, and they must be able to handle a higher peak-to-average power ratio (PAPR) than 802.11 PAs [8].

Second, phase noise must be considerably better than in an 802.11 design. Tighter phase noise requirements have implications for the synthesizer, which result in a longer settling time.

Third, if I/Q interface is chosen, then I/Q balance must be tighter as well, and will likely require I/Q calibration.

As we follow the design model of 802.11 for designing 802.16 Tx/Rx components, considerable effort must be made to develop higher efficiency, more linear PAs, and adaptive predistortion will be needed to achieve high linearity with high efficiency.

6.6.3   RF Architecture

When selecting radio frequency (RF) architecture for a WiMAX design, the basic choice is between a superheterodyne and direct-conversion architecture.

In terms of satisfying the transmitter regulatory requirements, a superheterodyne architecture is advantageous because of off-chip filtering of unwanted emissions but is expensive.

A direct-conversion transmitter architecture, on the contrary, is attractive because it leads to a smaller and less expensive radio design. The challenge with this approach is that performance is harder to maintain. For instance, any small DC offsets that occur will degrade system performance [8].

6.6.4   Chip Design Options

IC designers must decide how to partition functionality in a chip. While it is possible to integrate the Tx and Rx, with both the RF and intermediate frequency (IF) sections on one chip, this approach is not common today. For a superheterodyne architecture, it is common to partition the chip [8].

The partitioning can be done either at RF/IF stage where transmit and receive are on the same chip, but with separate chips for RF and IF. This is a better alternative as one synthesizer can be shared between both ICs [8].

The partitioning can also be done at Tx/Rx level. In this there is a separate Tx and Rx chip, and both include separate RF and IF chains.

To achieve best performance at lowest cost, IC makers can use different process technologies for the two ICs. For instance, it is possible to use a silicon CMOS process for the IF chip, and SiGe or GaAs for the RF device.

6.6.5   Regulatory Issues

The regulation agency is planning a procedure that allows a simplified and fast assignment of licenses, which can be limited to regional areas or to other technical parameters. WiMAX efforts are currently focused on test equipments supporting 256 OFDM physically operating in 2.5, 3.5, and 5.8 GHz frequency bands.

6.6.6   Cost

One challenge is the ability to deliver a well-priced product to end users. WiMAX shifts the cost curve of mobile data to compete on a cost-per-megabyte level [9] with cable and DSL, but the technical bugs will make deployment difficult in the beginning.

6.6.7   Spectrum Availability

Spectrum congestion is another issue for WiMAX, which is going forward. Spectrum needs to be available for WiMAX to be successful [9]. Wi-Fi uses spectrum in bands that are unlicensed in most regions of the world. The situation for WiMAX is much more complex because of the higher transmit-power levels and the fragmented radio spectrum in both licensed and unlicensed bands, which differ from country to country [10].

6.6.8   Spectrum Interference

WiMAX uses 5 MHz channels that sometimes splat all over Wi-Fi and interfere with WLAN and 3G bands. This issue is a challenge. Moreover in security and emergency services, there is no room for signal interference, which calls into question the viability of unlicensed spectrum, such as 2.4 and 5.8 GHz [7]. However, vendors are working on technological solutions (such as smart antennas and multiple-input, multiple-output [MIMO]) and the Federal Communications Commission (FCC) appears to be readying itself to expand available spectrum.

6.7   WiMAX as a Solution to 4G: Introduction

Mobile WiMAX is a 4G technology that is fairly well accepted and will offer broadband data, voice, and video services. At the moment, WiMAX technology will build upon Wi-Fi in a very complementary way, expanding open standard-based wireless networking for metropolitan area outdoor mobility applications including voice over IP (VoIP). Hence WiMAX with Wi-Fi can be called as a migration path to 4G [11]. While Wi-Fi is able to provide high-speed, localized, wireless Internet access, the emerging WiMAX standard is a wide area technology, supplying wireless coverage over an area of several kilometers. WiMAX (IEEE 802.16) with Wi-Fi (IEEE 802.11) will allow operators to deliver high-quality voice, video, and data services on a metropolitan scale and can provide users with connectivity wherever they are. Thus, WiMAX will deliver the promise of 4G, expanding open standard-based wireless networking for metropolitan area outdoor mobility applications.

This part provides a very strong comparison between WiMAX with other BWA 4G-based technologies, deployment of various architectures like (i) WiMAX micro-cells formed with Wi-Fi to provide micro mobility, (ii) WiMAX macro-cells that provide macro mobility, and (iii) mobile WiMAX architecture for full mobility.

6.8   Evolution of 4G

4G is basically a wireless phone standard and has a life of its own. It is a part of the wireless telephone family [12]. The original analog and digital cellular services were invented to cut the wire on landline phone service and provide a regular mobile telephone service. As such, the bandwidth they offer for adding data services is pretty meager, in the low Kbps region.

The wireless phone family started competing with the wireless Internet family. Now that a wireless/cell phone is not merely a cell phone, but also a personal digital assistant (PDA), a messaging system, a camera, an Internet browser, and an email reader. This is the place where the whole concept of "G" thing got started [12]. This new generation of cell phone service has been dubbed 3G for third generation. 3G has been proven to be a tough generation to launch. The demand for greater bandwidth right now has spawned intermediate generations called 2.5G and even 2.75G. While 3G phones and services are just starting to come into their own, the industry is developing triple play models, where voice, data, and video get bundled called 3¼G technologies such as high-speed downlink packet access (HSDPA). There is also an emerging cellular standard with true broadband data speeds (20 Mbps) called 4G. These speeds enable high-quality video transmission and rapid download of large music files.

While the story of wireless phone family is on one side, the wireless Internet family, which includes Wi-Fi and the emerging WiMAX standards, has started offering wireless VoIP telephone services, competing with 3G, and achieving the goals of 4G. These are completely different set of standards based on IP solutions. The first device-level WiMAX products, such as laptop cards, will take from 2 to 5 years to come and will enable a user to remain connected within an entire metropolitan area.

Recently, a product portfolio known as Wi4, which melds mobile WiMAX (802.16e) and 4G technologies, has been introduced by Motorola [13] that can deliver 300 Mbps in a fully mobile environment. Wi4 solutions will use all-IP access technology, peer-to-peer architecture, and "zero footprint" BS solution that will eliminate the need for a great deal of equipments used in legacy cellular networks. Also Korea's telecom industry has developed its own standard, WiBro, which is derivative of the mobile version of WiMAX [14].

So, a slugfest between cellular phone services and the BWA technologies offering VoIP services is going on. Cellular is trying to carve out its own unique markets next to wireless broadband. However, the question is whether wireless broadband currently like WiMAX be able to deliver 4G goals or will it have to merge with cellular to migrate toward 4G. But there are also possibilities that the two are likely to merge so that they can support both the cellular and VoIP standards.

6.9 Viability of WiMAX as an Alternative to 3G and Migration Toward 4G

Introduction of mobility into the WiMAX roadmap, as well as the slow and expensive rollout of 3G in many areas of the world, has changed the picture, and WiMAX is seen as a direct challenge by some 3G operators. Also, significant investments have been made by operators into Wi-Fi, which is likely to play an integral part in many eventual WiMAX deployments. This section gives the answer to the mobile carriers' burning question, how WiMAX will pose a significant threat to cellular voice and will compete with their cellular networks and migrate toward 4G.

1. Technology: The 3G cellular standards such as UMTS (Universal Mobile Telecommunications System), W-CDMA (wideband code division multiple access), and EVDO (evolution data only) are looking increasingly obsolete, well before 3G has been able to establish itself in the market in a serious way. These technologies are threatened by the shear superiority of BWA technologies such as WiMAX and FLASH-OFDM [15]. WiMAX is based on OFDM technology paired with MIMO smart antenna technology, which is best suitable for 4G.

2. Data speed: The limitations of W-CDMA, along with the fact that it requires capacity for cellular voice and data services, mean that operators are unlikely to support 3G connection speeds greater than 100 Kbps. EVDO has download speeds up to 2.4 Mbps. The HSDPA technology enables downlink with data transmission up to 8–10 Mbit/s. But WiMAX supports data downlink speeds in excess of 500 Kbps several kilometers from a BS.

3. Coverage range: The voice requirement limits the W-CDMA cell size, restricting coverage to just 2–3 km from a BS [15]. WiMAX has a range of 4–6 miles (30 miles max).

4. ROI: Mobile WiMAX will enable wireline telecommunication operators' and second tier operators with no 3G license to challenge the big cellular carriers and could disrupt the cellular operators' plans to gain ROI on those licenses and their investment in 3G infrastructure.

5. Spectrum: WiMAX's genuine advantage over 3G is that it can work in either licensed or unlicensed spectrum, whereas 3G cellular system requires licensed spectrum.

6. Interference [16]: WiMAX based on OFDM utilizes multiple channels to send and receive data, which results in less interference than 3G cellular data systems such as 1xEVDO and HSDPA.

7. IP connectivity: WiMAX supports ATM, IPv4, IPv6, Ethernet, and VLAN services. So, it can provide a rich choice of service possibilities to voice and data network service providers.

8. Interoperability: The architecture shall lend itself to integration with an existing IP operator core network (e.g., DSL, cable, or 3G) via interfaces that are IP-based and not operator-domain specific. This permits reuse of mobile client software across operator domains.

9. Cost of deployment: In addition to the advantages in converging mobility, portability, and fixed Internet access, the cost factor is also important for any deployment. WiMAX infrastructure is far cheaper than cellular.

10. Backhaul: WiMAX provides backhaul connections to cellular services.

11. Standardization: This means that the big mobile operators could potentially migrate their networks with mobile WiMAX technology, but since they have their own legacy networks you need to make sure you connect to them correctly. The WiMAX Forum works for standardization and will make sure multiple vendor solutions are available.

12. Economies of scale: At the moment, BWA vendors rely on proprietary, custom-built chipsets, which have kept equipment costs high and customers away. With standardization comes the ability to mass produce products, so research and development costs decline along with manufacturing expenses. This could pave the way for lower-cost services, which WiMAX proponents say will make BWA feasible for mass deployment.

6.10   Characteristics of 4G/WiMAX Network

4G wireless networks can be realized with an IP-based core network for global routing along with more customized local area radio access networks that support features such as dynamic hand-off and ad hoc routing as well as newer requirements such as self-organization, QoS, multicasting, content caching, etc. For a successful deployment of the 4G technology, it is imperative that we define the vision for the 4G services and applications that effectively meet the users need. The vision that is driving us is the users' vision as listed below:

1. Broadband: You want to be able to send and receive all kinds of information—images, video, big data files, and the same can be done at any location. 100 MHz per operator is required for 4G voice, video, and data services.

As applications require more data rates that is where technologies such as mobile WiMAX are valuable and the goal is to build bridges to traditional 3GPP and 3GPP2 standards.

2. Mobility: With 4G, a wireless customer might be able to take a car or train ride coast to coast and surf the Web uninterrupted.

With WiMAX network service providers will now enjoy the flexibility to market mobile, nomadic, and fixed services at true broadband speeds to customers with extremely efficient control over the access and use of their radio spectrum and network resources.

3. Roaming between various networks: If you are moving around and you go from a high-speed wireless LAN to a cell or satellite system, you want the hand-off to be smooth. Your applications should adapt gracefully, not choke or drop out. The key here will be enabling the "hand-off" procedures that allow a mobile device to switch the connection from one BS to another, from one 802 network type to another. For example, a notebook could connect via Ethernet or 802.11 when docked, and stay connected with 802.16 when roaming the city or suburbs.

The goal is to standardize the hand-off so that devices are inter-operable as they move from one network type to another. The aim is to build a network interface to these networks connecting to all networks including DSL and cable infrastructure because we want WiMAX technology to be adapted by both the mobile operators and fixed operators.

4. Convergence: You want to be able to access the network from lots of different platforms, such as cell phones, laptops, PDAs.

WiMAX is a powerful system that delivers connectivity intelligent and flexible enough to support streaming video, VoIP telephony, still or moving images, email, Web browsing, e-commerce, and location-based services through a wide variety of devices. That means freedom for consumers.

5. Efficient: In addition to being a lot more cost efficient, 4G is spectrally efficient, so carriers can do more with less. WiMAX based on OFDM is spectrally efficient. Also we believe that IP-enabled technology will provide a lower cost and faster time-to-market solution.

6. Access over inaccessible area: With wireless systems, mobile users can be contacted almost anywhere, anytime. The ability to quickly provision service, even in areas that are hard for wired infrastruc-ture to reach; for example, if rivers, free ways, or other obstacles separate buildings you want to connect, a WMAN solution may be much more economical than installing physical cable or leasing communications circuits.

7. Harmonization: As there is no single network blanketing the globe, but rather a vast patchwork of networks, the whole thing has to be

heterogeneous and seamless. So, the goal is to build a new high-capacity, high-quality interoperable broadband network that can carry any content that any consumer wants to have. That is where WiMAX comes in. However, since chipsets are custom-built for each BWA manufacturer, this adds time and cost to the process of bringing a product to market.

6.11   Impact of Similar Technologies on WiMAX

Unlike earlier BWA iterations, WiMAX is highly standardized. WiMAX will compete with W-CDMA and CDMA2000, as well as other technologies like HSDPA that support both voice and data services. UMTS is a direct competitor to WiMAX because it is also IP based. The "Super 3G" standard, which is based on an upgrade for W-CDMA and is reported to support speeds 10 times faster than UMTS, is also expected to compete with WiMAX. But mobile WiMAX is expected eventually to provide data rates of about 20 Mbps, although initial implementations may be closer to 1–2 Mbps and will overcome all similar technologies. Tables 6.1 and 6.2 provide a comparison of the speeds of delivering data and video and steaming audio of various technologies, respectively.

TABLE 6.1

Data Speed of Various Technologies

Technology	Bandwidth (MHz)	Max Speed to Deliver Data (Mbps)
W-CDMA HSDPA	5.00	14.4 (fixed and mobile)
CDMA2000 1xEVDO RA	1.25	3.2 (fixed and mobile)
IP Wireless (UMTS TDD)	5.00	7.5 (fixed and mobile)
Flarion	1.25	3.1 (fixed and mobile)
WiMAX	20.00	75 (fixed/nomadic)
	5.00	4–18 (fixed/nomadic)
Mobile WiMAX	5.00	Up to 15 (mobile)

Source: Seybold, A.M., Outlook 4Mobility 2Newsletter, 2004.

TABLE 6.2

Speed of Delivering Video and Steaming Audio

Technology	No. of Channels in 10 MHz	Total Max Speed to Deliver Video and Steaming Audio (Mbps)
W-CDMA HSDPA	2	28.8
CDMA2000 1xEVDO RA	7	31.1
IP Wireless (UMTS TDD)	2	15.0
Flarion OFDMA	7	22.4
WiMAX OFDMA	2	8–46
WiMAX Mobile OFDMA	2	Up to 30

Source: Seybold, A.M., Outlook 4Mobility 2Newsletter, 2004.

From the tables, it is clearly seen that mobile version of WiMAX reportedly supports high data rates and it promises to provide mobile support for speeds up to 150 km/h and hence provides a migration path to 4G.

6.12   VoIP Services of Wi-Fi/WiMAX

WiMAX is initially targeted at bridging and access applications for inter-networking or Internet access. But eventually mobile applications for voice, video, audio, and data were built in to the standard because it is acknowledged that data, by itself, do not provide a business model for WiMAX. Let us see the VoIP services offered by Wi-Fi and WiMAX.

6.12.1   Vo Wi-Fi

Wi-Fi phones are the next generation intelligent IP communications devices. They add session initiation protocol (SIP) or H.323 based VoIP communications together with Wi-Fi installations. These phones can be used in any Wi-Fi network [11]. There are also soft phones, which can be downloaded into a PDA or laptop with additional software that turn into wireless speakerphones when connected to Wi-Fi networks. Two protocols are currently being used, namely, the H.323 and SIP. At present, vendors are offering hard phones based on both H.323 and SIP.

6.12.2   Mobile WiMAX

The main focus of mobile WiMAX is targeting what is generally called 4G, which is still several years away [18] . Availability of cheap mobile (or at least portable) WiMAX clients is the first requirement to support mobility. Mesh networking is a way to route data, voice, and instructions between nodes. It allows for continuous connections and reconfiguration around blocked paths by hopping from node to node. Only a small amount of meshing is required to see a large improvement in the coverage of a single BS.

They are experimenting with Wi-Fi SIP phones and expect a transition at some point from Wi-Fi to WiMAX when mobile 802.16e handsets are ready.

6.12.3   Hybrid Cell Phones

Wi-Fi enabled cell phones that bring together cellular and landline phone systems. The CL400 is a camera/MP3 phone that offers GSM/GPRS and tri-band (850/1800/1900) capabilities for global operation with Wi-Fi (802.11 b/g) connectivity. In the future, the CL400 will also support voice over IP (VoIP), based on SIP standards.

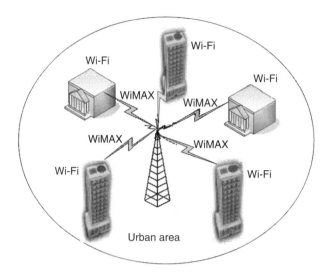

FIGURE 6.5
Wi-Fi cells embedded to form one WiMAX micro-cell.

6.13   Various Deployments Scenarios of WiMAX

We enter into 4G of networks by deploying architectures that is a mix of VoIP and mobile IP-based solutions. This will be based on WiMAX micro-cells formed with Wi-Fi supporting data and voice services that initially will be deployed for dense urban areas and will interconnect with WiMAX macro-cells through interoperable WiMAX PTP links. In the future, when mobile WiMAX enters the market it will be deployed in the similar fashion as cellular service fulfilling the goals of 4G.

6.13.1   WiMAX Micro-Cells Formed with Wi-Fi (Migration Path toward 4G)

At the moment businesses should consider using both Wi-Fi and WiMAX, because it will take a while for WiMAX radios to get inside laptops. Wi-Fi is already in laptops, cell phones, and PDAs. One of the first uses of WiMAX will be backhaul for Wi-Fi hot spots forming a micro-cell. With WiMAX, hot spots will be converted into so-called "hot zones," especially for users of Wi-Fi VoIP phones, such as the VoiceLine XJ100 by Net2Phone and hence leads the path to 4G. WiMAX may be used to tie together these municipal Wi-Fi mesh networks so that a VoIP phone user traveling from one place to another would never lose coverage. Though the WiMAX standard does not describe how much capacity an operator can feed each Wi-Fi access point, a single WiMAX BS could handle hundreds of megabits per second of data and can feed one or more Wi-Fi APs mounted on tall buildings (Figure 6.5).

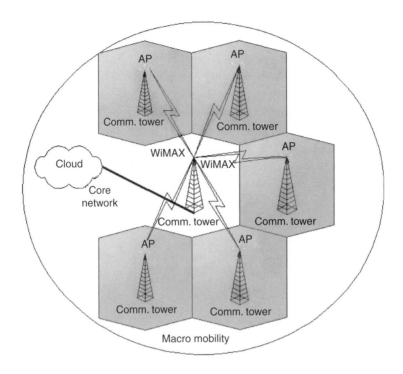

FIGURE 6.6
WiMAX cluster providing macro mobility.

Figure 6.6 shows WiMAX main BS with wired backhaul at the center of a cluster of WiMAX Mesh BS forming a macro-cell providing coverage for the surrounding region [11].

WiMAX mesh substations in turn can feed Wi-Fi hot spots in both WiMAX main or mesh cells as shown in Figure 6.5.

6.13.2 Deployment of Mobile WiMAX (Complete 4G)

Mobile WiMAX will be available beyond 2007–2008 and allow VoIP phone users to seamlessly communicate over metropolitan areas, large outdoor hot spots, or campuses served via WiMAX spectrum, especially in rural or hard-to-reach areas. Users will be able to stay best connected—connected by Wi-Fi (802.11) when they are within a hot spot, and then connected by 802.16 when they leave the hot spot but are within a WiMAX service area.

To offer mobility, the wireless service must be as pervasive as cell phone service. The solution is to create small cells instead of trying to cover large areas with a single antenna. Figure 6.7 shows how mobile WiMAX will provide full mobility like cellular.

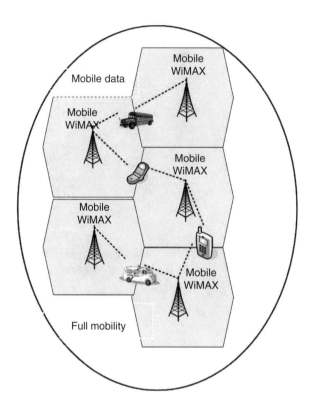

FIGURE 6.7
Full mobility by mobile WiMAX.

6.14   WiMAX Enables Rural Telecommunication Infrastructure: Introduction

The lack of access to reliable energy remains a significant barrier to sustainable socio-economic development in the world's poorest countries. Majority of their population are largely concentrated in the rural areas, and access to power is often sporadic or altogether lacking. Without power the traditional telecom infrastructure is impossible. So, if electricity is playing havoc, there is a need to devise low-tech solutions to help bridge not only the digital divide but also the electrical divide. One of such solutions is a solar- and pedal-powered remote ICT system by Inveneo a nonprofit organization, which combines the power of the computer and a clever application of the increasingly popular Wi-Fi wireless technology powered by solar energy. With this system the rural villagers pedal onto the hand-built, bicycle-powered PC in the village which would send signals, via an IEEE 802.11b connection, to a solar-powered mountaintop relay station. The signal would then bounce to a server in the nearest town with phone service and electricity and from there to the Internet and the world. In this part, we describe a prototype of how the wireless broadband WiMAX technology can be integrated into the existing system and gain global possibilities. With the new suggested prototype each village will connect to one WiMAX station through the Wi-Fi AP that is solar-powered. The WiMAX tower then sends the radio signal to fixed fiber backbone that will connect the villages to the Internet and enable VoIP communications.

6.15   The Suggested Remote ICT System with WiMAX

The low-cost solar- and bicycle-powered ICT system uses standard off-the-shelf PC, VoIP and Wi-Fi services, and open source software technologies that have been designed for low-power consumption and integrated, ruggedized, and adapted for the local environment and language.

The computer will be powered by electricity stored in a car battery charged by foot cranks. These are essentially bicycle wheels and pedals hooked to a small generator. The generator is connected to a car battery and the car battery is connected to the computer. The pedal-powered village PCs would interconnect between themselves on a WLAN and each PC in turn connects to an "AP" that relays message packets between different destinations. The AP is connected to the WiMAX relay station. With WiMAX we can reach about 50 km point-to-point broadband. The WiMAX tower in turn connects the villages to the other WiMAX tower or an AP connected to the fiber backbone through a server in the nearest town and from there to the Internet and the world. The AP is a solar-powered IEEE 802.11b (Wi-Fi) connection.

6.15.1   Description about the New System

The system is based upon low-power embedded PCs running the GNU/Linux operating system. The PC also supports two PCMCIA slots to accommodate an IEEE 802.11b WLAN card supporting Wi-Fi wireless communications and a VoIP card (H.323) supporting voice communications [19]. Phone CARD DSP/Phone interface card can use standard analog phone as well as headset/microphone combination. A cluster of PCs in a village uses Wi-Fi to send data wirelessly to a central WiMAX tower. A single WiMAX tower can serve many clusters as shown in Figure 6.8.

The system uses a pedal to charge a car battery, which is attached to a special cycle battery unit called RP2, which, in turn, is connected to the PC to run it in those locations that do not have power. RP2 is a power management system that switches the computer to a power battery when the power phases out. RP2 system can provide continuous power for about 8 h. Moreover, 1 min of pedaling yields 5 min of power. HCL Infosystems [20] has designed the prototype of an external gadget that can be charged through pedaling and connects

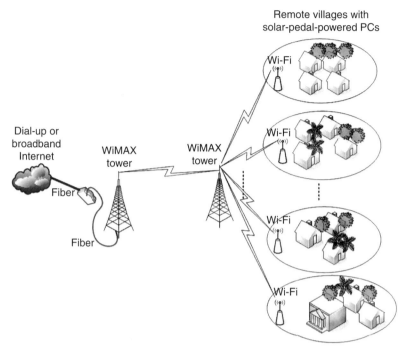

WiMAX with solar-powered Wi-Fi serving the pedal-powered PCs that
enables VoIP communication system of rural villages

FIGURE 6.8
Remote ICT system with WiMAX.

to a personal computer to run it under the most difficult power situations,
and this can be easily used. The system consists of four distinctive parts:

1. Main server: This system is placed in a location where phone lines,
   Internet access (dial-up or any kind of broadband), and electricity are available. The server incorporates a modem (V.34) and a
   PSTN interface card capable of emulating a telephone and converting the voice signals to and from digital form. The main server acts
   as follows:

   • It acts as the gateway to the local phone network (PSTN, analog
     or digital).

   • It maintains the connection to Internet (Internet access gateway—
     dial-up or broadband).

   • It handles the voicemail system with mailboxes for individuals.

   • It acts as the intranet web server for local content, file sharing,
     network monitoring, etc.

2. Relay system: This system consists of a WiMAX tower that acts as a repeater, extending the range of the signal from the main server to the AP and further toward the village PCs. The WiMAX tower can be solar powered if there is no provision of electricity within 2–6 km from the Wi-Fi APs. From the WiMAX tower a PTP connection can be established to another WiMAX tower in the city extending the range of the wireless network by relaying the signal from the village PCs to the server PCs in the city. A single WiMAX tower can feed several Wi-Fi APs within the region. Other features include:

   - Extends the range of village PCs from 50 km away to the main server.
   - Enables PTP or PMP connections.
   - Multiple relay stations can be connected to the central sever to cover large areas.

3. Solar-powered access point: 802.11 wireless network links act as the AP and ranges from 2 to 6 km. The PC is wired to a regular telephone set and a directional Wi-Fi antenna that transmits the Internet signal to the AP and routed to the WiMAX tower.

4. The village PC: This system provides the users with access to a phone line, email, web browsing, and basic computing applications. The village PCs are interconnected using wireless networking and have a telephone interface, so telephony is carried out using the standard telephone "human interface." Calls between villages and village clusters are routed by the router and cost nothing like dialing another room from a hotel PBX.

6.16   Advantages of the Remote ICT System with WiMAX

The use of the WiMAX technology in the ICT system has major advantages among the other ones.

- Practical limitations prevent cable and DSL technologies from reaching many potential broadband customers. WiMAX provided broadband services to rural areas in a cost effective manner.
- WiMAX also provides backhaul connections to the Internet for Wi-Fi hot spots in the remote locations.
- The network is designed and built in such a way that it will cost very less around $25 a month to operate.
- Even though the pedal device can also be powered by solar or gas generator, the idea is that young people will earn money/computer time pedaling the device.

- In addition to fulfilling the desire for telephone service, there are basic computer functionalities available for preparation of documents and spreadsheet functions.
- Because much of the project can be built around nonproprietary, or "open source," software, villagers can essentially own the system.

6.17   Significant Need for Remote ICT System with WiMAX

The following factors show the need for remote communication system:

1. Family communication: The global population shift from rural to urban communities inevitably breaks up families. These remote ICT networks allow distant family members to remain in contact, with positive results for community stability.

2. Health care: Heath clinics can communicate in real time with doctors and nurses in hospitals, provide AIDS awareness and prevention information [21], address complex medical treatment needs and emergencies, etc.

3. Human rights: Communities get access to information allowing them to take part in shaping their own destiny. They share information on human rights, women's rights, and land issues, improving farming techniques, etc.

4. Education: The integration of ICT in teaching curriculums increases availability of literacy and other training, provides youth opportunity to acquire computer skills, etc. [21].

5. Economic empowerment: Beyond the support for traditional economic practices, the introduction of information, communication, and energy technologies allows for the development of useful trade skills related to those technologies, from solar technicians to software programming.

6. Disaster relief: Rapid deployment of phone and data networks after disasters.

7. Income generation [22]: Through improved communication, farmers access market data to maximize crop value by taking it to the highest paying nearby markets. Coops are formed between villages to improve buying power and share resources. This results in substantial increase in income.

8. Aids distribution: Access to databases in real time provides resource information on grants and funding from government agencies and NGO entities.

9. Communication and transportation [21]: Improves local communication via phone and email—eliminates time and expense to make the full day journey between villages.

6.18 General Suggestions

- To get interoperability between products and services initially, WiMAX Forum must narrow down the specifications to a certain set of system options.

- BWA reliability is the most important selection criterion for service providers. The service providers' survey rated reliability even higher than cost. The WiMAX Forum must see that early adopters of 802.16 are satisfied with the technology's reliability as they are satisfied with 802.11.

- Flexibility must be a prime consideration when it comes to selecting an architecture that supports WiMAX not only as it stands today but on how it will evolve in the future.

- Municipal governments must adopt strategies regarding the development of regional WiMAX networks and providing ubiquitous coverage of high-speed Internet access to residential and business customers throughout their regions stimulating regional economic development.

- Several conferences and seminars must be organized to examine the pros and cons of building an in-house product and review the technical abilities necessary to succeed.

- 3G and WiMAX should be seen as competitive or complementary and business models that will benefit the operators must be derived.

- To make strategic investments in a few companies that plan to demonstrate how WiMAX leads to 4G and can be put to profitable use.

- To develop handsets and software that will allow roaming between different networks and Wi-Fi, WiMAX, and 3G cellular connections together can achieve the goals of 4G.

- To develop intelligent, programmable edge devices so that the impedance mismatches between systems and technologies can be handled. This is an important aspect of 4G.

- As a telecommunication system it is obvious that long service life would be important for the remote ICT system and so the network design must accommodate it.

- The remote ICT system had to be made as automatic as possible and simple enough to be operated by villagers to reduce operating costs.

6.19   Conclusion

Broadband is growing, but in many regions of the world, mostly in developing nations large swathes of the population has no access to DSL, cable modems, or even basic Internet access. WiMAX will play a pivotal role to solve the above problem by extending the growth cycle of broadband technology and providing affordable broadband for all and improve the quality of life.

In 5 years, 4G should be a reality and VoIP is going to drive new revenues and business models in the decade to come. Certainly, analysis done in this chapter shows that WiMAX will win in the marketplace and will capitalize on this and further supports the proliferation of VoIP devices and IP-based services. To stay competitive, carriers must think of offering both WiMAX and 3G. Otherwise they will loose the business in the future when 3G becomes obsolete and outdated. At the moment, combination of WiMAX with Wi-Fi has the potential to compete on a cost-per-megabyte level with cable and DSL. The main advantage of this combination is that the whole geographic area becomes a hot zone and leads the path toward 4G. In the future, mobile WiMAX is expected to offer full mobility broadband for all type of services (data, voice, and video) and deliver the goals of 4G technology.

Moreover with WiMAX, new VoIP infrastructure can be easily put up in rural communities, like the one suggested in Sections 6.7 through 6.13, which can help sending two-way voice signals with computers, mimicking the traditional phone system, and can make a big difference to the people in the rural areas. So each country belonging to the third world must adopt this system, which is cost effective, and must improve the living conditions of the people, which in turn will lead, to economic empowerment. Many companies, such as HCL Infosystems [20] of India, manufacture the new affordable model that is charged by pedal power that can be adopted. But government and other aid agencies must develop policies to implement the communication infrastructure with WiMAX, so that rural areas are easily connected. I hope that this system will soon become ubiquitous in the poor parts of the world and transform the third world.

References

1. J. Ghering, *WiMAX to dominate BWA*, Daily Wireless, 21 April 2004. Available online: http://www.dailywireless.com
2. X. Song, *WiMAX: broadband wireless access*, 24 September 2004. Available online: http://www.wi-fiplanet.com/tutorials/article.php/3412391
3. WiMAX Forum, *Promoting interoperability standards for broadband wireless access*, WiMAX Forum, 2004. Available online: www.wimaxforum.org
4. D. J. Johnston and M. La Brecque, *IEEE 802.16 WirelessMAN specification accelerates wireless broadband access*, 28 July 2003. Technology @ Intel Magazine. Available online: http://www.intel.com/update/contents/st08031.htm

5. M. Paolini, *Wi-Fi, WiMAX and 802.20—The disruptive potential of wireless broadband*, Senza Fili Consulting & BWES Ltd, March 2004.
6. Intel, *IEEE 802.16 and WiMAX-broadband wireless access to everyone*, Whitepaper, 2003. Available online: www.ieee802.org/16
7. Pyramid Research, *Is surveillance a WiMAX killer app*, 17 July 2004.
8. D. Poulin, WiMAX *advantages bring about new challenges*, Commsdesign, 25 August 2005. Available online: http://www.commsdesign.com/article id=170100112
9. T. Seals, *WiMAXimum exposure—A new type of broadband wireless gathers momentum*, Infrastructure Solutions, January 2004. Available online: http://www.x-changemag.com
10. P. Fuller, *WiMAX opens wide range of design options*, Wireless Net DesignLine, Bath, England, 14 March, 2005. Available online: http://www.wirelessnet designline.com
11. V. Gunasekaran and F. C. Harmantzis, *Migration to 4G-ubiquitous broadband-economic modeling of Wi-Fi with WiMAX*, Proceedings of WWC, May 2005. Stevens Institute of Technology, USA.
12. J. Shelper, *1G, 2G, 3G, 4G*, TechColumn, April 2005. Available online: www.T1Rex.com
13. D. Jackson, *Motorola announces plans to converge WiMAX and 4G*, Primedia, Inc., July 2005.
14. J. Walko, *Samsung demos Korean version of WiMAX at 4G Forum*, Personal Tech pipeline eNewsletter, CMP Media LLC, August 2005.
15. Blue print, *The next bout: 3G versus BWA*, Available online: http://www.arcchart.com/blueprint
16. Mobile Pipeline Staff, *Wireless broadband called long-term winner over 3G: Study*, Mobile Pipeline eNewsletter, CMP Media LLC, February 2005.
17. A. M. Seybold, *WiMAX again?*, Outlook 4Mobility 2Newsletter, 2004. Available online: www.outlook4mobility.com
18. S. Buckley, *WiMAX Forum gears up for mobility and more*, Telecommunications Americas, Horizon House Publications, Inc., Gale Group, 2005.
19. C. Liddell, *Pedal powered: Look ma no wires*, Australia internet, 2002. Available online: http://www.australia.internet.com
20. M. Bakshi Chatterjee, *Bridging digital divide—New pedal power to run your computers*, Business Line, 2005. Available online: http://thehindubusinessline.com/2005/07/29/stories
21. N-TEN, *Inveneo-solar/pedal powered ICT*, Tech Success Story, 2005. Available online: http://www.nten.org/techsuccess-inveneo
22. Inveneo, *Pedal and solar powered PC and communications system*, 2005. Available online: http://www.Invenoe.org

7

WiMAX over GSM for Basic IP Access in African Rural Areas

Damien Chatelain and Barend J. van Wyk

CONTENTS

7.1   Introduction

The development of telecommunications in Africa has shown tremendous growth over the past few years. However, this development has not been homogeneous. Many major cities in Africa have services equivalent to those found in Asia or Europe. Global system for mobile communications (GSM), general packet radio service (GPRS), universal mobile telecommunications systems (UMTS), as well as asymmetric digital subscriber line (ADSL), and wireless IP networks are available. These islands of technology on the African continent are separated by thousands of kilometers of under-developed terrain encompassing mountains, grasslands, farms, villages, and deserts, where there is minimal or nonexistent electrical, road, or fixed telecom infrastructure [1]. Most operators are willing to bring telephone and Internet access to these disadvantaged populations, but they are con-fronted with severe constraints. Some of these are: a lack of basic supporting infrastructure; the theft of material; low densities of customers; potentially low return on investment; and a nonuniform distribution of potential sub-scribers (i.e., an isolated farm or lodge can potentially generate more income for a cellular operator than a complete village). Despite these obstacles, many rural areas in South Africa and Africa have GSM access. In many cases, networks in these areas are over dimensioned and operators are trying to introduce new services to improve their profitability. On the con-trary, potential customers like schools, Internet cafes, village councils, small companies, farms, and dispensaries simply need basic and cost-effective access.

Wireless solutions, especially Wireless Fidelity (Wi-Fi, IEEE 802.11) and WiMAX (IEEE 802.16), have been successfully implemented in low-income areas. In Indonesia, such networks have been deployed on a dollar per user basis [2]. Wi-Fi transmitters and other required components can be acquired on an astonishingly low budget [3]. WiMAX, which is still rela-tively expensive today, but should be more affordable in the near future, opens up even more promising opportunities for operators. In terms of cost and reliability, it is on par with ADSL [4], but connecting to a reliable core network remains problematic in African rural areas, unless the GSM network can be used for this purpose. In Africa, GSM covers areas where fixed telephony and often electricity are not commonly present [1]. How-ever, traffic per cell is generally low and the utilization of the E1 time slots on the Abis interface between the base transceiver station (BTS) and the base station controller (BSC) is still lower. The purpose of this chapter is to show that basic but affordable IP connectivity can be provided to rural communities by using spare capacity on GSM networks to carry WiMAX traffic.

In Section 7.2, a survey of different access technologies, with an emphasis on the cost for the user, cost for the operator, and rural deployment, is given. In Section 7.3, different techniques for integrating wireless and cellular net-works are investigated. Section 7.4 presents an architecture where GSM is used as a core network for WiMAX, the detail of an associated new proce-dure, and the results of tests with a South African operator. Conclusions are presented in Section 7.5 and some recommendations for future work can be found in Section 7.6.

7.2   Data Access Survey

Since 3G is already present in South Africa and Mauritius and is expected to be tested in Northern Africa by the end of 2006, it is natural to ask why WLAN over GSM is under consideration. Mainly on the basis of enhanced data rates for GSM evolution (EDGE) technology, 3G currently offers a maximum downlink speed of 384 Kbps and a maximum uplink speed of 64 Kbps. Although the system provides mobility, it is mainly rolled out in main cities and has limited prospects of getting deployed in rural areas [5]. The main reasons are the huge capital investment required by operators to upgrade 2G to 3G and the resulting high monthly subscription fees [5]. Moreover, surveys conducted by Vodafone in the United Kingdom show that voice contributes 80% of the overall revenue structure of this service, with SMS second at 16%, and data at only 4% [6]. Why implement such a service in rural areas where GSM traffic per cell is already very low?

In contrast, GPRS is widely deployed. It is present in nearly all areas where there is GSM, reaching 90% of the territory and 95% of the population in South Africa. In other African countries, between 20% and 90% of the territory is covered, corresponding to most of the urban and suburban areas [7]. Although operators in Africa are reluctant to invest in areas where no profit can be expected without government support, GSM has penetrated the African continent more than any other type of telecommunication technology. For the operator, upgrading from GSM to GPRS does not imply significant changes in the access network. The investment is therefore minimal compared to 3G rollout. This service, providing a maximum rate of 44 Kbps on the downlink, costs between 20 cents and 2 U.S. dollars per Mbytes. The subscriber mainly pays for mobility, which is not useful in cases where the service is used to connect a fixed PC to the Internet. If the use of GSM in rural areas is low, it is logical that the use of GPRS is even lower owing to more expensive handsets, modems, and connection fees, placing it out of reach of the majority of the African rural population.

Wireless data networks (wide area networks and WLAN) on the basis of the IEEE 802.11 and IEEE 802.16 standards are currently perhaps the most promising wireless technology. Given their popularity in developed countries, it is reasonable to consider the use of wireless access in developing countries as well. The forces driving the standardization and proliferation of WLAN in the developed world could also stimulate communications market dynamics in the developing world [8,9]. Such wireless technologies can easily be adopted as possible solutions for rural and under-served areas by virtue of ease of set-up, use, and maintenance, their relatively high bandwidth, and most importantly, their relatively low cost for both users and operators [9]. The radio part of this system is cost-effective, but an operator has nevertheless to link it to the rest of a network, which is not only expensive but also problematic. Data network solutions such as *ad hoc* and mesh networks [10] seem promising in terms of cost, flexibility of implementation, voice services, and even mobility with the coming 4G [11], but rollout in rural communities will be unlikely when GPRS, which is widely available in such areas, is still underutilized.

Since the main problem with WLAN in Africa is not the last mile, but rather finding a way to connect the wireless access point to an existing backbone network, a solution to integrate WiMAX/Wi-Fi with GSM is proposed. This solution would not only boost the usage of GSM access networks in rural areas, but also enable potential subscribers to gain access to IP services at reduced rates. Moreover, the implementation requires a relatively small investment and it could potentially allow a GSM operator to migrate directly from 2G to 4G in rural areas, thus avoiding the expenses associated with 3G rollout.

7.3   Different Ways of Integrating a Wireless Network with a Cellular Network

There are essentially two different methods for merging WLAN and cellular networks, namely, loose coupling and tight coupling [12].

7.3.1   Loose Coupling

As shown in Figure 7.1, loose coupling implies merging at the core network level. The cellular and WLAN access are independent, and the IP backbone is simply linked to the cellular core network. Therefore, the relations between the networks are minimal. Such a type of network has been successfully implemented by Nokia [13]. It allows sharing part of the core network and functionalities like authentication, locating, or charging. It also permits roaming and the use of most of the services offered by the GSM operator. Dual-purpose WLAN and GSM enabled seamless use of mobile phones on both networks.

7.3.2   Tight Coupling

As shown in Figure 7.2, for tight coupling, merging of the networks is performed at the access level. In this solution, a unique core network is shared. The WLAN access point can be connected to the BSC in the case of GSM, or to a radio network controller (RNC) in the case of UMTS [12]. A BSC/wireless interface is used, and the deployment of a large number of small units is preferable to avoid having long transmission links.

Nevertheless, tight coupling necessitates a new access network. For rural areas such transmission links used for access are expensive to install and (at least in the short term) will not be optimally utilized. Despite these difficulties, the GSM Abis interface can be used for this purpose. A later section will show how to optimize the use of the Abis interface by introducing WLAN access at the BTS level.

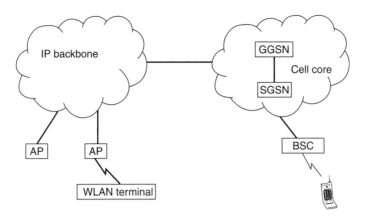

FIGURE 7.1
Loose coupling.

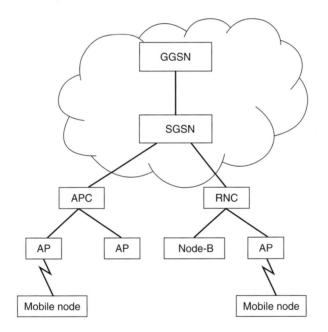

FIGURE 7.2
Tight coupling.

7.4 WLAN over GSM

The minimum bandwidth of the Abis interface is 2 Mbps. In many cases (at least in South Africa), operators use cascaded BTS configurations in semirural and rural areas to provide several times the capacity of an E1. This capacity

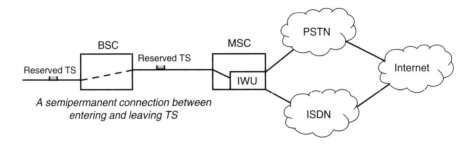

FIGURE 7.3
Modifying GSM network.

is dedicated to GSM traffic, which can be quite low in many rural areas. The integration of IP traffic on the Abis interface, providing capacity to a WLAN network, is a way to improve the utilization of the GSM network in these areas. Two possibilities have been investigated, namely, sharing the Abis interface and overlapping WLAN (based on WiMAX or Wi-Fi) and GSM traffic.

7.4.1 Sharing the Abis Interface

To interface a WLAN with the GSM network, several modifications have been implemented. It seems possible to interface WLAN with GSM without acting on the BSC or mobile switching center (MSC), but in such a case, the interface has to comply strictly with GSM standards, which complicates implementation. With only a few modifications to the GSM network, a much simpler solution can be designed. The WLAN can be interfaced with the discontinuous cross connector (DXX), and the BTS configuration does not need to be modified. As shown in Figure 7.3, a semipermanent connection is established through the BSC and MSC. This connection through the BSC implies that a time slot is reserved before and after the BSC, and that the only task of the BSC is to copy the time slot received to the time slot transmitted.

A direct connection can be established between the WLAN interface and the interworking unit (IWU) in the MSC. The IWU can be connected to the Internet through the fixed network. Its function is, for example, to convert cellular traffic to Internet traffic [14]. If such a system is used, the WLAN interface is directly connected to the IWU by means of reserved time slots and then the WLAN interface can establish a connection to the Internet. Instead of having to respect GSM norms to communicate with the BTS, the WLAN interface simply has to communicate with IWU, which is much easier.

7.4.1.1 *Modifying a Linux Kernel to Offer a Transparent Link to Applications and Routing*

As explained before, the WLAN interface is directly connected to IWU and has to communicate with it. A new protocol has been developed for this purpose. Its function is to establish a connection to the IWU and to organize IP data to make it readable by the IWU. As shown in Figure 7.4, this protocol is

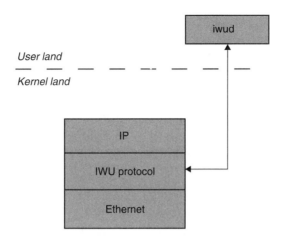

FIGURE 7.4
Adding a new protocol.

localized between IP and Ethernet on the OSI layer scheme. Indeed, Ethernet is the data link protocol. This interface is connected to DXX using a single Ethernet card. A Linux operation system (OS) is used. The implementation of this new protocol, called the IWU protocol in the sequel, is divided into several parts, which are discussed below.

- First, the main functions of this protocol have to be implemented in the kernel domain.
- Then, in the user domain, a daemon called iwud is implemented to take care of signaling, for example, to establish or to terminate a connection.
- Finally, this protocol should be transparent to applications. By implication, it should be transparent to load sharing or quality of service (QoS) software.

On a Linux system, the only way applications can communicate with the kernel is by using the system call *socketcall*. Although this system call is almost never used directly by programmers, libraries have been developed. Most C programmers, for instance, use the BSD socket library.

A C structure named the *net device* can be used to implement this interface. The kernel domain provides an interface to the middle of OSI layer 2, since programming under this layer depends on the hardware used, and so kernel developers have defined a software interface to allow hardware industries to create drivers for the Linux kernel easily. When a packet (*skb* structure in Linux; the *skb* structure is associated with a packet) is processed by the IP layer, the IP protocol calls *skb-¿dev-¿dev queue xmit*, where *dev* is the *net device* structure, representing the interface where the packet has to be sent according to the routing table. This means that, if a new interface (called IWU) is set up, the IP protocol will call it if the routing table shows that this kind of packet must be forwarded to the IWU (more information can be found in Ref. 15). At this layer, the routing capabilities of Linux and applications can be used. The IWU is transparent but is not the final destination.

The IWU interface is implemented as a tunnel interface since it is not linked to specific hardware. After a packet has been processed by the IWU protocol, it is forwarded thanks to a classical *dev queue xmit* call by the Ethernet interface.

In this case, the achieved throughput is proportional to the number of Abis time slots (64 Kbps) allocated to the WLAN. However, this configuration is not ideal since a logical link, and hence permanent bandwidth, is reserved. It provides a new service but does not improve the utilization of the existing infrastructure.

7.4.1.2   Use of Discontinuous Cross Connectors

In some cases, operators connect GSM equipment to the DXX to create several logical networks on the same physical one. Therefore, the DXX interface can be directly connected to an IP network and act as a router. In this way, access to the Internet can easily be granted and existing billing facilities can be set up.

7.4.2   Overlapping WLAN and GSM Traffic

7.4.2.1   Quasi Best-Effort Service

GPRS offers some QoS mechanisms [16]. It is accordingly possible to have a simple low-cost solution, not providing strict best-effort treatment but a compromise solution by making use of GPRS QoS mechanisms. GPRS provides for different classes of services, namely, precedence class, delay class, reliability class, and throughput class [16], defining a QoS profile.

When a GPRS connection is set up, a QoS profile is negotiated. A QoS profile implies that the attributes of a connection (delay, reliability, precedence, mean, and peak throughput) are associated with a specific class. Practically, for this service, the lowest QoS profile will be requested:

- Precedence class number 3
- Best-effort delay class (delay class number 4)
- Reliability class number 3
- Mean throughput class number 1 (which means best-effort)
- Peak throughput class number 1

To avoid a situation where the serving GPRS support node (SGSN) refuses the profile, a service access point identifier (SAPI) with the lowest QoS (unacknowledged mode) has been defined in the logical link control (LLC) layer. The SGSN then has to be configured to associate this profile with this best-effort SAPI (as shown in Figure 7.5).

The SAPI used for this QoS profile has to be specified to avoid any compression and to minimize the use of SGSN processing resources. This configuration is not perfect, since nothing is implemented to prevent the

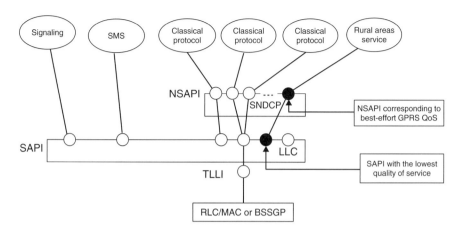

FIGURE 7.5
Using SAPI to create a best-effort service.

SGSN to multiplex another packet data protocol (PDP) on the best-effort SAPI. The fact that the SGSN multiplexes a PDP on the best-effort SAPI means that the other PDP is of the same priority as the rural service (requested the same QoS profile), and therefore it is not seen as an important problem. The defined mechanism allows the SGSN to treat the data when it has spare resources [17] and is therefore a good compromise.

7.4.2.2   Best-Effort Traffic on the Gb Interface

As seen in the previous section, best-effort QoS treatment in the SGSN does not result in a full-fledged best-effort service, since WLAN data is only processed when spare resources (processing power) are available. Alternatives for creating a full-fledged best-effort service are consequently investigated. This section describes a solution using the Gb interface.

Sometimes layer 2 of the Gb interface is provided by a frame relay operator. For this reason, acting on the traffic has to be done just after SGSN or BSC. For implementation, a few modifications have to be made on the BSC and on the SGSN in the network service layer [18]. From the network service layer point of view, the connection between a BSC and SGSN is identified by a network service virtual connection identifier (NSVCI), which maps a frame relay permanent virtual connection (PVC).

The idea is to multiplex two NSVCs on the same PVC. Figure 7.6 presents the details. Two classical routers are used as multiplexers. These multiplexers allow one NSVC to have priority over the second one, which therefore becomes best effort from a network point of view.

Between BSC and SGSN, a new NSVC is defined. The data of the service is routed to this new NSVC, which fits into a PVC connected to a frame relay router. PVCs between the SGSN and the router, and between the BSC and

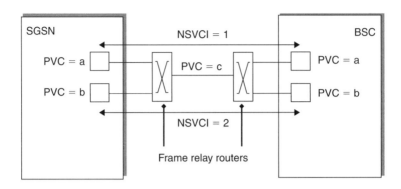

FIGURE 7.6
Multiplexing two network service virtual connections on one permanent virtual circuit.

router, are configured, and on this part of the network there is reservation of resources. However, these routers have to be located close to the SGSN and BSC, typically in the same room, and so the network is oversized on this part (for tests a new PCM trunk was added). Frame relay equipment normally offers some QoS mechanisms (like using committed information rate) but these are too basic to create the desired best-effort service [19]. Current routers, like the ones used for tests and configured for frame relay transmission, allow for powerful QoS mechanisms (priority queuing, for instance). With equipments meeting these specifications and the proposed architecture, the network will give priority to the classical GPRS traffic on the Gb interface.

7.4.2.3 Multiplexing of GSM/GPRS and Ethernet Traffic on the Abis Interface

To combine GSM traffic with Ethernet traffic (to be used for WiMAX), an E1/Ethernet multiplexer and a new procedure were developed. The way this works is set out below.

For the E1 input:

- If a time slot is used for synchronization (TS0) or signaling, no multiplexing.
- If a time slot is used for traffic, multiplexing on the time slot is allowed.

For the Ethernet input:

The Ethernet packets are sent to the transmission buffer. When an Ethernet packet has to be transmitted, it is divided in four-byte blocks, which should be forwarded on the same E1 output time slot. If there is an error in the transmission of the Ethernet packet, the Ethernet packet is sent to the retransmission buffer. A counter is used to verify that the transmission of each Ethernet packet is complete. For each four-byte block, an identity byte can be added to maintain the correct order of transmission and to rebuild the Ethernet packet at the receiving side.

For multiplexing, each traffic time slot of the E1 input is monitored. Multiplexing is therefore performed on a time slot by time slot basis. Since the E1 TS of the Abis interface is divided into four traffic channels, the multiplexing process has to check four times whether the Abis time slot to be used is empty. For these reasons, the Ethernet frame is also transmitted 4 bytes at a time and, if an error occurs during transmission, the full Ethernet packet is sent to the retransmission buffer to be transmitted again. Note that the acknowledgement of the transmission is not done at this level, but only at the Ethernet protocol level.

The complete proposed architecture with the protocol stack of the Ethernet transmission is shown in Figures 7.7 and 7.8.

7.4.2.4 Simulations, Implementation, and Tests of the Overlapping WLAN and GSM Traffic Solution

The simulations and the tests of the new architecture have been performed using Matlab®, Simulink® (Figure 7.9), OMNET C++®, Cisco® routers, and the developed multiplexer. An OMNET C++ simulation of the complete architecture was done before implementation and testing in The French South African Technical Institute in Electronics (F'SATIE) telecommunications laboratory.

To implement the algorithm in a real network and to verify the accuracy of the proposed algorithm, two test multiplexers/demultiplexers were implemented using field programmable gate array (FPGA) technology with four E1 ports and the Ethernet port on the input, and four E1 ports as outputs. The complete state diagram for multiplexing is given in Figure 7.10.

In the state diagram the abbreviations noted below are used.

- Sync: Synchronization of the E1 line with the time slot 0
- Send: Check if Ethernet packets are received on the Ethernet port
- TxD: Transmission data to the E1 output port
- RxD: Reception data on the E1 input port
- ZD: Zero detector
- Addr: Address
- Txcnt: Transmission counter (2 bit counter)
- Rxcnt: Reception counter (2 bit counter)
- PZD: Previous zero detector
- L: Counter of byte for the Ethernet transmission
- Len: Length of the Ethernet packet
- Ercnt: Error counter
- Start: Check if data are received on the demultiplexer

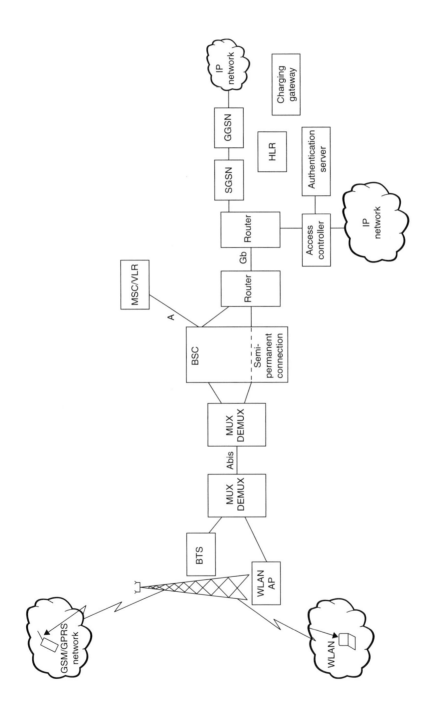

FIGURE 7.7
Description of the complete proposed architecture.

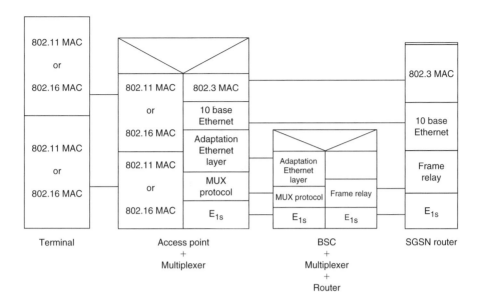

FIGURE 7.8
Protocol stack of the Ethernet transmission over the GSM network.

An explanation of the different states follows.

- IDLE: It synchronizes with the E1 line. If nothing is received on the Ethernet port, the data coming from the E1 input port are transmitted to the E1 output port. As an exception, if the value ⟨1111 1111⟩, corresponding to the hexadecimal value ⟨F F⟩ is received, the value ⟨1111 1110⟩ corresponding to the hexadecimal value ⟨F E⟩ is transmitted.

- SEND: It is used when Ethernet packets have to be transmitted (send=1). The zero detector checks if the E1 time slot is empty during four consecutive frames. As there are four traffic channels per time slot on the GSM Abis interface, the zero detector checks whether each of the traffic channels are empty. If there is traffic on the E1 input port, this traffic is sent to the E1 output port. If there is no traffic on the E1 input port, the value ⟨F F⟩ is transmitted on the E1 output port and the transmission counter txcnt is set to zero.

- TX4Z, WRITE, and READ: These three states allow the transmission of 4 bytes of an Ethernet packet on the E1 output port. The counter txcnt is incremented for each transmission of a byte. When it reaches the value ⟨11⟩, corresponding to txcnt(2)='1', the next state diagram is entered.

- INCRL: It checks if there is still nothing received on the E1 input port with the previous zero detector. PZD periodically checks the

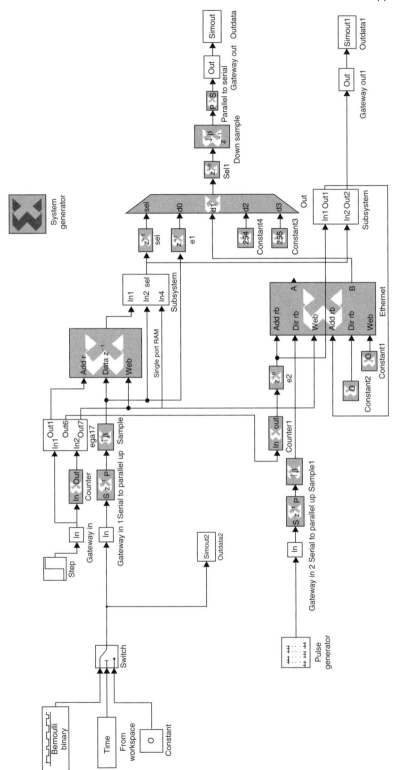

FIGURE 7.9
Simulink model of the E1/Ethernet multiplexer.

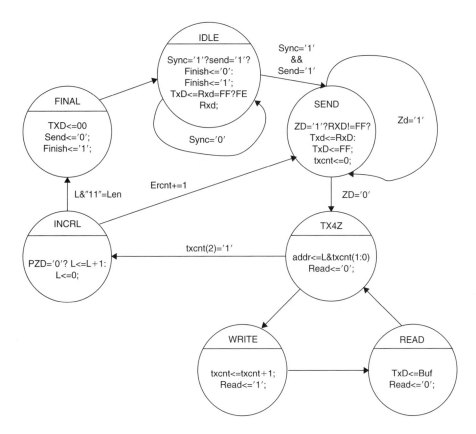

FIGURE 7.10
Multiplexing state diagram.

content of each of the four traffic channels allocated to one E1 time slot in the Abis interface. If there is no transmission on the E1, L is incremented and the transmission of the Ethernet packet continues on. Moreover, L is set to zero, the Ethernet packet has to be retransmitted, and Ercount counter is incremented. If the value of Ercount reaches a preset value (4 for example), the Ethernet packet is forwarded to another E1 time slot. If the value of Ercount stays more than 0 for a set period of time (30 ms for example), the Ethernet packet is also sent to another E1 time slot.

- FINAL: This state is reached when an Ethernet packet has been completely transmitted (L&"11"=Len). It allows the transmission of Ethernet packets with a fixed length by using an Ethernet bridge at the Ethernet input and by setting Len to a specific value, or for transmitting Ethernet packets with variable length. The Ethernet bridge preserves the processing capacity of the FPGA. It can facilitate the hardware implementation and decreases the retransmission rate, which occurs when large Ethernet packets are transmitted.

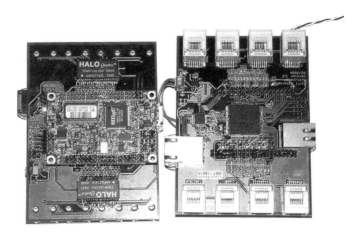

FIGURE 7.11
Hardware of the Ethernet/E1 multiplexer.

In fact, the size of an Ethernet packet is variable and always a multiple of 4 bytes. When the packet is completely transmitted, the byte ⟨0 0⟩ is sent to indicate the end of an Ethernet packet to the receiver.

The multiplexer presented in Figure 7.11 was tested on one time slot using an E1 tester to generate traffic for the E1 input port, an Ethernet router to generate Ethernet traffic, and a digital analyzer to verify the multiplexing and demultiplexing process.

The tests focused on three critical aspects for the multiplexing and demultiplexing are:

- The beginning of the transmission of an Ethernet packet
- The end of the transmission of an Ethernet packet
- The E1 traffic occurring during the transmission of an Ethernet packet

The complete architecture has also been simulated with OMNET++, as shown in Figure 7.12. The GSM and Ethernet data coming from the WiMAX are collected on the BTS and forwarded through the multiplexer. Data are delivered via E1 links to the demultiplexer of the BSC. The Ethernet data exit the BSC via a semipermanent connection, acting as a transparent link to connect a Cisco router. Frame relay over E1 connectivity is established between the two routers as on the Gb interface of the GPRS network. The subsequent tasks are divided into three steps.

First, frame relay connectivity between the two routers is established. Next, the multiplexers are linked to the routers with E1 and provide priority to GSM/GPRS traffic. Finally, GSM and Ethernet traffic is supplied to the routers (Figure 7.13).

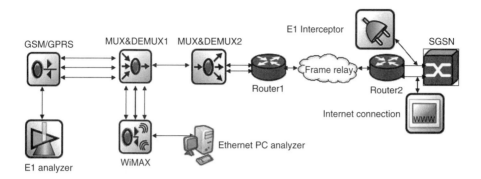

FIGURE 7.12
OMNET++® model of the complete proposed architecture.

FIGURE 7.13
Implementation of the proposed architecture.

7.4.2.5 Quality of E1 Traffic and Ethernet Throughput

The quality of E1 traffic (voice and data) and Ethernet throughput has been analyzed for the following three cases:

- E1 voice integrated service digital network (ISDN) traffic
- E1 GSM Abis voice traffic
- E1 GPRS Abis data traffic

7.4.2.5.1 E1 ISDN Voice Traffic

Even if the multiplexer has not been primarily designed for voice ISDN traffic, testing has revealed that it could be used with the both A-law and Mu-law coding. The bytes ⟨1111 1111⟩ and ⟨1111 1110⟩ are quantization steps apart corresponding to numerical values +126, +127 (A-law) and 0, +1 (Mu-law). The value ⟨1111 1111⟩ is reserved by the multiplexer; therefore, voice is coded with only 255 steps, instead of 256, causing a degradation of quality. Fortunately, this degradation is not perceived by the ear and is similar to when conversion from A-law to Mu-law takes place between Europe and the United States, creating a distortion in the companding level.

For a standard ISDN configuration, the maximum latency for the multiplexer to switch from Ethernet to voice traffic is 1.125 ms, corresponding to a loss of 9 bytes. Every time the multiplexer switches the mode of transmission, a maximum of 1.125 ms of conversation is lost. This is still acceptable since it corresponds to the latency of a normal silence detector [20]. It even allows transmission during a conversation with the use of a speech detector, if time assignment speech interpolation (TASI) [21] is not implemented or in the case of low traffic.

The Ethernet throughput that can be achieved is 51 Kbps per time slot in the absence of voice traffic, and up to 25 Kbps per time slot when a conversation occurs. A complete E1 connection (with one time slot dedicated to synchronization and one time slot dedicated to signaling) could thus carry up to 1.5 Mbps when not used for conversations, and approximately half this value otherwise.

7.4.2.5.2   E1 GSM Abis Voice Traffic

Since the coding of voice (260 bits every 20 ms) and the configuration of the E1 line (four traffic channels per time slot) are different on the Abis interface than for ISDN, the results change as well. Since the multiplexer changes the state ⟨1111 1111⟩ to ⟨1111 1110⟩, it creates an average increase of the bit error rate (BER) by 0.05%. Knowing that standard vocoders accept BER of higher than 1%, in some cases up to 4% [22,23], the BER generated by the multiplexer is negligible in comparison with the BER created by the interference on the air interface, which cannot be corrected by the channel coding.

In this case, the maximum latency of the multiplexer to switch to the E1 traffic is 3 ms, with a loss of 2 bytes. In the worst case, where these 2 bytes cannot be recovered by the vocoder, a loss of 20 ms of speech can be expected. According to Ref. 21, a user does not perceive a loss of speech of up to 300 ms, and is not much affected by a loss of speech of up to 600 ms. Moreover, during handover in the GSM network, it is common to lose several nonperceivable bursts of speech.

In terms of throughput, the results per time slot are similar to the previous section. However, the standard configuration of an Abis interface allocates three time slots per transmitter: two for traffic and one for the signaling. In order not to disturb the network, no multiplexing should take place on the signaling channel. Therefore, the maximum Ethernet throughput, which is possible to achieve if the E1 is not used, is 1 Mbps. Fifty percent of the capacity dedicated to the Abis interface could be reused for data transmission.

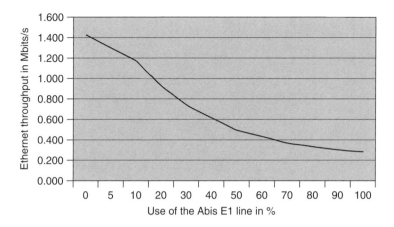

FIGURE 7.14
Ethernet throughput versus use of Abis interface (GSM traffic, LAPD concentrated, 4 transmitters per cell).

Finally, most of the operators use the discontinuous speech transmission (DST) feature to reduce the power consumption of the transmitters and interferences in the radio part. Fifty to sixty percent of a conversation consists of silent periods, which are not transmitted when using DST [21]. The multiplexer, with DST enabled on the GSM network, delivers a throughput of 0.063 Mbps per E1, when voice traffic occurs on all the traffic channels, and 0.25 Mbps when half the traffic channels are busy, assuming four traffic channels per time slot and normal configuration of the Abis interface (three Abis time slots per transmitter). In South African rural areas, traffic peaks at 20% of the available capacity during peak hour. According to Figure 7.14, representing the Ethernet throughput with link access procedure on the D-channel (LAPD) concentrated configuration of the Abis interface and four transmitters per cell (the optimal configuration for the use of the multiplexer corresponding to 2.25 Abis time slots per transmitter), the multiplexer can provide 0.5 to 1.3 Mbits/s per E1 to such areas.

7.4.2.5.3 E1 GPRS Abis Data Traffic

The transmission of data on the Abis interface uses the radio link control (RLC)/medium access control (MAC) structure described in Ref. 18. Data are naturally more sensitive to transmission errors than voice [24]. The block error rate (BLER) generated by transmission errors causes block retransmission and decreases the throughput. The BLER is linked to the coding scheme (CS) and the BER after decoding. The higher the CS, the lower the capacity dedicated to data coding is and the higher the sensitivity of the data to interferences [25]. An example of the direct relationship between the BER after decoding, the BLER, and the CS, is given in Figure 7.15 [26].

In the multiplexer, the BER created by the change of ⟨1111 1111⟩ to ⟨1111 1110⟩ is increased by 0.05%, which corresponds to an extra BLER of

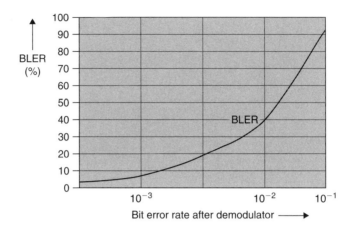

FIGURE 7.15
Relationship between BLER and BER after demodulator for coding scheme 1.

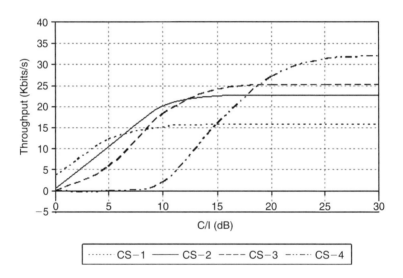

FIGURE 7.16
Link bit rates for different C/I ratios and coding schemes for the multiplexer.

2% for CS1 and up to 20% for CS4. The BLER and the throughput a user can expect are represented in Figure 7.16.

From simulations it is evident that the degradation in the transmission created by the multiplexer can only be tolerated when CS1 and CS2, which have smaller radio blocks, are used. The level of retransmission for CS3 and CS4 generated by the multiplexer creates such a significant loss of throughput that implementation is not recommended for these cases. Moreover, every time a radio block is sent during an Ethernet transmission, it has to be retransmitted because of the latency of the multiplexer.

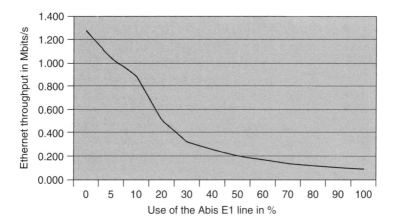

FIGURE 7.17
Ethernet throughput versus use of Abis interface (GSM + GPRS traffic, LAPD concentrated Abis, 2 transmitters per cell).

For the transmission of Ethernet packets, the throughput is lower when GPRS traffic is used instead of GSM. This is owing to the fact that virtually no Ethernet packets can be transmitted when a packet data channel is used on an Abis time slot. Stringent access control will also be necessary for Ethernet users to ensure noncongestion, since Ethernet traffic can be completely stopped if GPRS is used for more than 25% of the available capacity on the Abis link. Fortunately, standard GPRS configuration guidelines prescribe no more than four air interface time slots per user and eight air interface time slots per cell [27]. The simulated Ethernet throughput for a typical rural area configuration, with two transmitters per cell and four on demand GPRS time slots with LAPD concentrated Abis configuration, is represented in Figure 7.17. This configuration ensures that the Ethernet traffic will not be blocked.

7.5 Conclusion

The purpose of this work has been to provide WiMAX access to rural areas by using spare capacity on already existing GSM access networks. The literature survey showed that such an implementation has clear cost advantages. Traditionally, the integration of cellular and WLAN networks takes place in the switching system part. This study proposed integration at a lower level, utilizing unused capacity on the Abis links of the GSM network, to provide basic services at an affordable cost.

Naturally, the bottleneck created by the Abis interface is a limitation. This interface only provides several Mbps of capacity, whereas a WiMAX or Wi-Fi system can deliver more than 10 Mbps per user. However, this work demonstrated that by introducing a new procedure, it was possible to transmit

WLAN traffic on the Abis interface by adding a multiplexer developed and patented for this purpose, and that up to 70% of the Abis capacity could be reallocated for Ethernet traffic (carrying services like Internet). For a typical Abis interface, with four E1s correctly configured, up to 5.6 Mbps could be shared between different users. It is not enough to provide broadband services, but compared to <22 Kbps commonly available for a fixed line dial-up in South Africa, it could allow for the simultaneous connection of 250 users. 20 Kbps are enough for checking emails, to connect to Internet, and even to run a VoIP service like Skype®. The potential of such services at localized rural points like Internet cafes, city councils, hospitals, or isolated farms remains important if the costs remain affordable.

For an operator, the cost of this integration is typically 5000$ per router, 1000$ per multiplexer/demultiplexer, 150$ for a Wi-Fi access point, and 2500$ for a WiMAX access point. In most cases, routers are already available at the BSC level and the cost for WiMAX is continuously decreasing. It is even possible to integrate WiMAX with the multiplexer to reduce expenses.

For the user, the cost will be mainly linked to the tariff proposed by the operator. Wi-Fi or WiMAX cards are affordable for government or private institutions or Internet cafés. Since the operator reuses existing infrastructure, it should be possible to provide cost-effective IP access at a competitive price, when compared to alternative solutions.

7.6 Recommendations

By combining the solution proposed in Ref. 28 and the solution proposed here using tight coupling, 4G services could be delivered.

The integration of the GPRS interworking function (GIF) at the BSC level, as well as the WLAN adaptation function (WAF) as shown in Figure 7.18, could allow a customer to use WLAN network when it is available and the GSM/GPRS otherwise. The mobile station (MS) then supports two radio subsystems for transporting GPRS signaling and user data. The first interface is implemented with the GPRS-specific RLC/MAC and physical layers, whereas the second is implemented with the 802.11/802.16-specific MAC and physical layers. These two interfaces provide two alternative means for transporting LLC packet data units (PDUs). Typically, when the MS is outside a WLAN area, LLC PDUs are transmitted over the GPRS interface (Um). However, when the mobile enters a WLAN area, LLC PDUs are transmitted over the WLAN interface. This switching is performed with the aid of WAF and could be completely transparent to the user and to upper GPRS layers.

This kind of architecture could allow basic 4G services. VoIP as well as low-rate video streaming could be achieved with basic infrastructure and the networks already in existence. Moreover, it could be a way for local operators, which cannot afford wideband code division multiple access

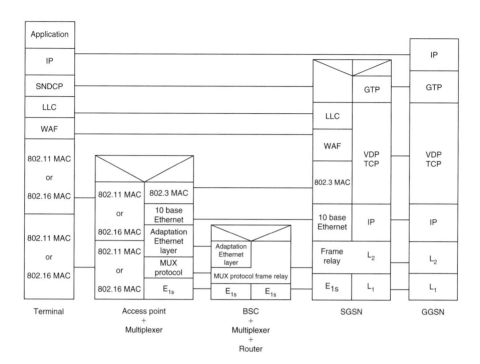

FIGURE 7.18
Protocol stack for combined cellular, WLAN application.

(W-CDMA), to provide simple and cost-effective services to disadvantaged populations.

Acknowledgments

The authors would like to thank the Centre of Excellence in Planning and Modeling, supported by Telkom, Alcatel, and Molapo for their financial support. MTN and Ericsson South Africa are also thanked for allowing the authors to perform tests in their laboratories.

References

1. L. Harris, Crossing a yawning chasm, *South African Wireless Communications*, vol. 9, pp. 18–22, September/October 2004.
2. O. W. Purbo, *Practical Guide to Build a Wi-Fi Infrastructure*, Computer Network Research Group, Inter University Center on Microelectronics, Institute of Technology Bandung, Bandung 40132. Available from onno-itb@kesemek.cs.ui.ac.id (Accessed on February 2004).

3. M. Outmesguine, *Wi-Fi Toys: 17 Cool Wireless Projects for Home, Office and Entertainment*, Indianapolis, IN: Wiley, 2004.
4. S. Chandler, D. Browne-Marke, and J. Redwood-Sawyerr, Development of WiMAX rural telecommunications system in Sierra Leone, *IEEE ICTe Africa 2006*, Nairobi, Kenya, 2006.
5. J. Harrison, *The 3G Report. Find Out if 3G is All It's Made Out to Be?*, Available from http://www.techdirect.co.za/3G.htm (Accessed on January 2004).
6. D. Crotz, Convergence—Can technology deliver? *Southern African Telecommunication Networks and Applications Conference (SATNAC) 2005*, Central Drakensberg, KwaZulu-Natal, South Africa (CD-ROM), 2005.
7. Available from www.cellular-news.com/coverage (Accessed on January 2005).
8. A. Pentland, R. Fletcher, and A. A. Hasson, *A Road to Universal Broadband Connectivity*, MIT Media Laboratory. Available from http://www.itu.int/ council/wsis (Accessed on March 2005).
9. F. Ohrtman, *WiMAX Handbook, Building 802.16 Wireless Network*, New York: McGraw-Hill Communication, 2005.
10. S. Basagni, M. Conti, S. Giordano, and I. Stojmenovic, *Mobile Ad Hoc Networking*, West Sussex, Great Britain: Wiley & Sons, IEEE Press, 2004.
11. A. J. Cole, Wireless carrier reform is essential for data services success, *IEC 2004 Annual Review of Communications*, vol. 57, section 1, pp. 55–60, 2004.
12. IEEE Standard. IEEE standard for Wireless LAN Medium Access Control (MAC) and Physical Layer (PHY) specifications, ISO/IEC 8802-11: 1999(e), August 1999.
13. J. Ala-Laurila, J. Mikkonen, and J. Rinnemaa, Wireless LAN, access network architecture for mobile operators, *IEEE Communications Magazine*, vol. 39, no. 11, pp. 82–89, November 2001.
14. Ericsson, *Understanding Telecommunications*, Part A: Chapter 9, 2004. Available from http://www.ericsson.com/support/telecom/part-a/a-9-3.shtml#marker= 61838 (Accessed on January 2005).
15. K. Wehrle, F. Pahlke, H. Ritter, D. Muller, and M. Bechler, *Linux Network Architecture*, New York: Prentice Hall, Chapter 14, pp. 263–314, 2004.
16. GSM 03.60. 3GPP Technical Specification, *Group Services and System Aspects; General Packet Radio Service (GPRS); Service Description*, Stage 2. Version 6.11.0, release 1997.
17. ETSI. Digital cellular telecommunications system (phase 2+), *General Packet Radio Service (GPRS); Sub-Network Dependant Convergence Protocol (SNDCP)*, ETSI GSM 04.65 GPRS standard, release 1998.
18. ETSI. Digital cellular telecommunications system (phase 2+), *General Packet Radio Service (GPRS); Base Station System (BSS)—Serving GPRS Support Node (SGSN) Interface; Network Service*, ETSI GSM 08.16 GPRS standard, release 1998.
19. V. Kosonen, *QoS and Frame Relay*, Helsinki University of Technology, Laboratory of Telecommunications Technology, HUT, Finland, 2003. Available from http://www.netlab.hut.fi/opetus/s38130/k99/presentations/12.pdf (Accessed on January 2005).
20. B. Waggener, *Pulse Code Modulation Techniques with Applications in Communications and Data Recording*, New York: International Thomson Publishing Inc., 1995.
21. Members of the technical staff Bell Telephone Laboratories, *Transmission Systems for Communication*, 5th ed., United States of America, Bell Laboratories Edition, 1982.

22. ETSI Standard GSM 06.60, *Digital Cellular Telecommunications System; Enhanced Full Rate (EFR) Speech Transcoding*, release March 1997.
23. S. Wanstedt, J. Petterson, T. Xiangchun, and G. Heikkila, *Development of an Objective Speech Quality Measurement Model for AMR Codec*, AWARE, Lulea, Sweden, Ericsson Research, 2002. Available from http://wireless.feld.cvut.cz/mesaqin2002/full12.pdf (Accessed on April 2006).
24. T. Halonen, J. Romero, and J. Melero, *GSM, GPRS and EDGE Performance*, 2nd ed., West Sussex, Great Britain: Wiley & Sons, 2004.
25. TR 45.050. 3GPP Technical Specification Group GSM/EDGE Radio Access Network, *Background for Radio Frequency Requirements*, release 7, 2005.
26. GSM 04.60. 3GPP Technical Specification, *Radio Access Network; General Packet Radio Service (GPRS); Mobile Station (MS)—Base Station System (BSS) Interface; Radio Link Control/Medium Access Control (RLC/MAC) Protocol*, Version 7.9.0, release 1998.
27. R. Alexander, Link-quality estimation method and components for multi-user wireless communication systems, United States Patent 6879813, 2005.
28. N. Passas and A. K. Salkintzis, WLAN/3G integration for next-generation heterogeneous mobile data networks, *Wireless Communications & Mobile Computing Magazine*, vol. 5, no. 6, pp. 599–601, September 2005.

8

Applications of Wi-Fi/WiMAX Technologies in the Emerging World

Vinoth Gunasekaran and Fotios C. Harmantzis

CONTENTS

8.1 Introduction

8.1.1 Wireless Broadband Revolution

The period of the Industrial Revolution during the past two centuries saw the most development in the history of mankind. But that period of unparalleled growth will be overshadowed by the current technological revolution, namely, the ICT revolution. This revolution will not only benefit individuals, but also enable tremendous impact on a nation and the global economy as a whole. Because of the increased global connectivity, the amount of information that can be transmitted electronically has grown exponentially, resulting in an unprecedented level of ease in communication between all parts of the world. For true success of the ICT revolution, one needs broadband connectivity, as it is not only the information that is shared among people but also the voice, image, video, etc. There is no agreed definition for broadband. It is usually recognized with higher transmission speeds and "always on" connectivity. Broadband is at the heart of the convergence of telecommunication, information technology, and broadcasting [5]. Therefore, there is a great need for modern hi-tech communication infrastructure, since the focus of applications is on interaction rather than just information sharing.

Compared to wireless technologies, wired networks give the same level of connectivity in some selected places, but they lack ubiquity and affordability. Wireless networks can be deployed much faster with less initial investment; they also offer a lot more flexibility in terms of adapting to changing bandwidth requirements. Although the cost has come down, the cost of civil engineering, site acquisition, and laying fiber or copper cables remains very high. It is essential to conduct cost/benefit analyses of deployment of wireless versus wired networks [8]. Wireless networks are easy to deploy and the service can be provided within days. As proposed by Pentland [14], wireless technology will be the first viable infrastructure to serve rural and underdeveloped areas. The rationale behind this assertion is that after the invention of the telephone, it took nearly 100 years for wired telephones to reach a population of one billion people around the world. But with the invention of cellular communication, it just took about 20 years to reach the same number.

Wireless broadband boasts of some big benefits over wired broadband networks. In the United States, some cities have started their initial phase of deploying citywide wireless networks, making ubiquitous broadband a reality. In some cases, the city's goals are just to improve the overall efficiency of the government services and deliver low-cost fixed broadband wireless Internet services to low-income communities and small businesses. Affordable wireless broadband access has the power to transform an emerging economy by inducing investment and innovation in e-commerce, e-education, telecommuting, e-health, agriculture, e-entertainment, e-government, and almost every other activity of the economy as discussed by Brewer et al. [2]. Most importantly, the Internet is on its way to become a day-to-day utility, where an affordable and ubiquitous broadband wireless access will be seen as an extension of everyday life. As seen in Figure 8.1, one has an option to migrate from narrow band and leapfrog to fixed wireless, instead of deploying wired broadband.

8.1.1.1 *Wi-Fi or Wireless Fidelity*

Wi-Fi or Wireless Fidelity permits connectivity to the Internet from virtually anywhere, at speeds of up to 54 Mbps. Wi-Fi enabled devices use radio technologies based on the IEEE 802.11 standard to communicate data anywhere within the range of an access point. There are three main issues in bridging the digital divide: affordability, availability, and accessibility of services and applications. Many technologies have been successful so far; however, most

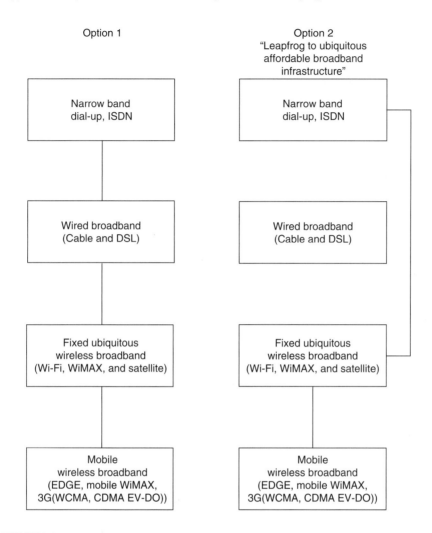

FIGURE 8.1
Internet access evolution and options for deploying Internet access.

of them fail to address key challenges in affordability, deployment, power consumption, etc. Wi-Fi is a technology that has the potential to address all these issues.

Wi-Fi accessibility: For most developing countries, lack of access to advanced voice and data services remains a barrier to network readiness. It is clear that within the next few years those who do not have access to the next generation of broadband-driven communications technologies such as VoIP (voice over Internet protocol), video telephony, and IPTV (Internet protocol television) will be at a great disadvantage [12]. The developing countries need to build a broadband communication infrastructure that is accessible to all, encouraging social services and e-government applications.

Wi-Fi wireless access technology is undoubtedly an attractive option for data, voice (i.e., VoWi-Fi), and video, as compared to other traditional communication infrastructures in the developing world.

Wi-Fi availability: In most countries, 2.4 GHz band is license-exempt, although some may require registration of use. Wi-Fi has become the most common use of unlicensed band for a "hot spot" or "hot zone" or "hot city" type of coverage. This is because of the widespread availability of Wi-Fi radios complying with IEEE 802.11b and the upcoming 802.11g standard. Wi-Fi has 100% global recognition and has become the single networking standard for all developers, equipment manufacturers, service providers, and end users. The main advantage with Wi-Fi is that a large-scale service-level roaming between different Wi-Fi providers is possible as Wi-Fi certification has become a *de-facto* standard for IEEE 802.11-based products.

Wi-Fi affordability: The benefit of using Wi-Fi in the "last mile" is that the client device is extremely inexpensive, owing to the large volume of production. Capital investment is also cost-effective, providing more flexibility over traditional wired communications which, in turn, results in low prices for Wi-Fi broadband services [14]. The standardization and interoperability between different vendor products made Wi-Fi prices very low and facilitated rapid penetration from a niche to the mass market around the globe. It is expected that at some stage, WiMAX will also reach a price and performance level similar to Wi-Fi. At least for the next few years, Wi-Fi will proliferate rapidly as a last-mile option and deliver wireless broadband access at a price dramatically lower than WiMAX.

8.1.1.2 WiMAX

WiMAX stands for worldwide interoperability for microwave access. IEEE standard 802.16 is the foundation of WMAN (wireless metropolitan area network) of the next few decades [7]. A group of vendors and service providers (those who founded the WiMAX Forum)* believe that WiMAX will be widely deployed in a similar manner to that of Wi-Fi. Standardization will not only reduce equipment and component costs, allowing mass production, but also allow interoperability between equipments of different vendors. The most suitable frequency band for WiMAX would be the 3.5 GHz band, followed by the 5.2–5.8 GHz band. It is also expected that a 2.5–2.7 GHz band would be a potential band for WiMAX in some countries. Technically, there are lot of differences between Wi-Fi and WiMAX as shown in Table 8.1. Broadly speaking there are only two types of Wi-Fi deployment: selected indoor locations or hot spots and extensive or outdoor coverage [3]. But on the contrary, there are several ways WiMAX can be deployed. The first type is the most popular one, providing backhaul for Wi-Fi access points (APs), and it can also serve

---

* www.WiMAXforum.org: The WiMAX Forum is an industry-led, nonprofit corporation formed to promote and certify compatibility and interoperability of broadband wireless products. Their member companies support the industry-wide acceptance of the IEEE 802.16 and ETSI HIPERMAN wireless MAN standards.

TABLE 8.1

Wi-Fi versus WiMAX

Characteristic	Wi-Fi	WiMAX
Spectrum	Unlicensed 2.4 and 5.8 GHz	Both licensed and unlicensed Spectrum 2–11 GHz
Coverage	(i) Designed for 300 ft range for indoor use	(i) Designed for 10 km; maximum range up to 30 miles.
	(ii) Due to recent innovations, coverage is being extended using a mesh technique or high-gain directional antennas for outdoor usage.	(ii) Basically designed for outdoor environments (terrains, buildings, trees, etc.)
Capacity	54 Mbps in 20 MHz channel	75–100 Mbps (based on a modulation technique and it is adaptive)
Channel width	200 MHz (fixed channel size)	3–20 MHz (flexible channel size)
Quality of service	VoWi-Fi is emerging and 802.11e a proposed standard by IEEE to define QoS in WLANS	Standard has inbuilt QoS for voice and multimedia applications

as a backhaul between conventional cellular towers. The second type would be as the last mile, which serves the residential and enterprise users as an alternative to cable and DSL [6,19]. The third type is similar to metro Ethernet provided on point-to-multipoint (PMP) sources that have direct competition with fiber. The fourth type is the mobile version of the WiMAX based on the 802.16e standard; although ratified recently, it is not expected to be quickly adopted by operators.

8.2 Broadband Infrastructure for Emerging Countries: The Case of India

India is the second most populous country in the world with an annual GDP per capita of around $1,000 whereas the United States, Japan, and Western Europe have GDPs of nearly $36,000 per year.* Although India's GDP per capita is very low, the growth rate of its economy makes the country attractive for foreign investments. India has been following a market-oriented approach and has also liberalized its economy. Its development has followed a process of "soft" industrialization in the past decade, focusing primarily on the services sector. Unlike other countries, India's development relies primarily on software and Information Technology services, which is, by far, the fastest-growing sector in its economy. One of the major factors contributing to India's growth is the large pool of skilled manpower. India has a lot of potential and

---

* http://www.cia.gov/cia/publications/factbook/docs/profileguide.html

opportunity yet to be explored as shown by Prahalad [15]. Focusing mainly on IT services, India is now entering an important new phase in its economic evolution.

India's goal is to become the engineering and knowledge process outsourcing hub of the world, after it has shown significant success in IT services and business process outsourcing. Since the future economy is knowledge-driven it is essential to make citizens' access to computers and the Internet—and, thus, the world—easier. Community tele-center projects were at the top of the developmental hierarchy in India. This strategy has to be taken further, keeping in view the fact that ICT will constitute the main tool to facilitate India's emergence as a leading knowledge society in the world. However, to build a strong civil society in a knowledge-based economy, a free flow of information in all tiers of the economy is needed. Hence, armed with an ubiquitous and affordable broadband infrastructure, India can lay the groundwork in its goal of equipping every citizen to face the challenges of an emerging knowledge-oriented economy.

As described by Moss et al. [13], telecommunications is the critical infrastructure for the twenty-first century just as highways were central to the twentieth century. This is because highways filled with cars and trucks have been superseded by the instantaneous, virtual data and voice freeways. Although India has entered the race for infrastructural development in the late twentieth century, it is now in the best position to adapt ICT-centric infrastructure faster than other developing countries. India has realized that ICT projects are the only enablers of social and economic development for the underserved regions. There are many successful ICT projects that are benefiting to ordinary Indian masses.* There are initial projects going on in India where the primary goal is to test, explore, and define the vision that ICT could be of constructive use to the local community, and, especially to a financially underprivileged community. In this market, ICT infrastructure backed by broadband connectivity can be viewed as both a social and an economic enabler.

8.3 Wireless Infrastructures for Urban and Suburban Areas

The modern communication infrastructure uniquely varies, not only from country to country but also within different parts of the same country. It is not necessary to have uniform infrastructure in cities and villages, since the needs and requirements of these two are different. It is important, however, to understand each wireless technology's capabilities and limitations and it is very much essential to figure out different architectures for each tier of

---

* By Kenneth Keniston, Grossroots ICT Projects of India, Available online at http://web.mit. edu/~kken/Public/PAPERS/ASCI_Journals_Intro_ASCI-version_.html (Accessed on May 2007).

the economy. It is also critical to sketch the wireless applications that can be offered in the near future, as new applications are expected to evolve over the next few years. Therefore, in deploying a wireless broadband communication infrastructure, the countries/states need to have both tactical and strategic vision. A citywide Wi-Fi/WiMAX deployment is emerging as a modern high-tech economic development tool. It is currently being used in developed countries, such as the United States, and can also be used in developing nations. In designing network architecture, it is necessary to find the social and economic value of the network. It is essential that wireless access should be a fundamental means for public communication in the near future. Each kind of wireless infrastructure has its own social and economic value [9,11].

8.3.1   Ubiquitous Wireless Infrastructure for Urban Areas

This model we consider incorporates *infrastructure WiMAX mesh* with *Wi-Fi systems*, where both technologies coexist to offer a cost-effective solution, as shown in Figure 8.2. Each metro zone has a WiMAX base station (BS) and it serves as a first, second, or third tier backhaul for all Wi-Fi mesh nodes within its coverage zone. The leasing cost should be low, as this infrastructure can use the light pole or the rooftop of a campus building, reducing operating costs significantly. The infrastructure WiMAX mesh architecture can cover the whole urban area by forming a number of metro zones. Infrastructure WiMAX mesh is a type of mesh where the subscriber nodes do not forward packets. Unlike other mesh networks, the infrastructure WiMAX mesh network is slightly different. It is a type of mesh that is different than "ad hoc" or "client" mesh. WiMAX main BSs having wired backhaul can be placed at the center of a cluster and Wi-Fi mesh APs are embedded in both WiMAX main and mesh cells. This is because the rental of the wired backhaul networks accounts for a major cash outflow [1]. The operating expenditure can be significantly reduced by using infrastructure WiMAX for backhauling [10]. Under each WiMAX cell coverage area, mesh Wi-Fi nodes can be deployed by giving blanket coverage. Roughly 20–40 Wi-Fi APs can be placed inside each WiMAX cell. Later on, as mobile WiMAX (based on IEEE 802.16e) becomes a reality, it can be substituted for last-mile access instead of Wi-Fi nodes. It can use the same backhaul infrastructure and the last-mile WiMAX migration would be much simpler in the later stage. The technology of wireless mesh routing is nothing but chaining together separate nodes. It is being viewed as a low-cost method for providing instant access to thousands of users.

   Mesh Wi-Fi in the last mile: IEEE is currently establishing a new standard called 802.11s* to extend mobility to Wi-Fi access points within very large Wi-Fi networks. There is also a lot of research going on with wireless LAN MAC and physical (PHY) layer for ESS (extended service set) mesh networking [4]. The Wi-Fi access can provide high bandwidth for very low

---

* IEEE website: http://grouper.ieee.org/groups/802/11/

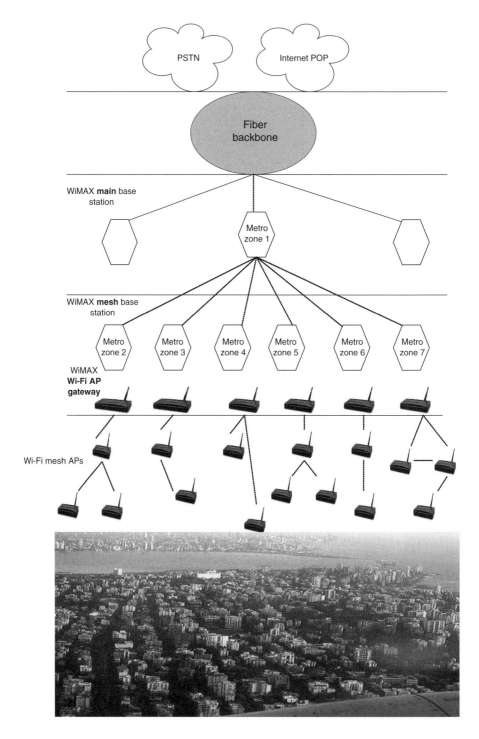

FIGURE 8.2
Urban infrastructure: Wi-Fi access in the last-mile "hot city." (Google Images.)

capital expenditure. Customers can get on-demand services and also monthly subscription plans. With portable devices having Wi-Fi clients, people can connect to the network while roaming in different metro zones throughout the city. Another main advantage of using Wi-Fi in the last mile is the various technology options it has. The Wi-Fi chip makers have already announced a tri-mode chip having IEEE 802.11b/g and 802.11a as their portfolio product.* The service providers can strategically plan to deploy their access points to support as many technologies and standards as possible. This would allow the client's software to sniff and select the best technology available at any given spot. The most widely known Wi-Fi standard 802.11b supports a smaller number of audio streams when compared to the high-performing standard 802.11a or 802.11g. Nevertheless, 802.11a with eight channels can be a technology of choice for voice applications, making it an attractive alternative to 802.11g, which has only three nonoverlapping channels. The Wi-Fi service providers can think of installing access points that include 802.11a for voice users and 802.11b for data users. Another way of taking advantage of these various technology options would be using 802.11a for Wi-Fi mesh links and 802.11b or g for Wi-Fi APs communicating with the client device. The urban metro zones can be formed in the following places:

Urban metro zones

- Corporate campuses
- Urban shopping centers and complexes
- Municipal/corporation offices
- Recreational area (parks, beaches, etc.)
- Airports, seaports, train stations, and bus terminals
- Residential area (high rise buildings)
- Universities, colleges, and schools
- IT parks, manufacturing and research parks
- Government buildings and campuses

8.3.2 Affordable Fixed Wireless Infrastructure for Suburban Areas (Hot Zones)

Wireless infrastructure ranges from less coverage configurations for rural areas to ubiquitous coverage for metropolitan areas. However, it is imperative to find an optimal coverage area for semipopulated areas. In that case, the capital expenditure is very low with WiMAX used as backhaul and Wi-Fi for access, covering a particular geographic zone. The coverage can be provided by Wi-Fi APs with high-gain antennae to extend the neighborhood coverage or campus area coverage. By taking advantage of various wireless

---

* http://www.intel.com/

technologies ranging from Wi-Fi long-range APs to wide-area WiMAX under one umbrella, residents and businesses of a suburban region are able to get wireless access. The broadband penetration in suburban India is very low and many small businesses and low-income households cannot afford leased lines. In some situations, the connection speeds are also very low, insufficient to support many applications.

Neighborhood area network: This kind of network deployment is centered on the notion of physical communities and neighborhoods, as discussed by Rao [16]. A neighborhood-owned Wi-Fi distribution point can facilitate sharing of infrastructure, by subscribing members in their own neighborhood or community. In some cases, one of the residents owns the Wi-Fi access point and opens up the network and spares bandwidth to everyone around his/her neighborhood. In this network type, individual buildings and houses may be packed closely to each other. A form of free or low-cost Wi-Fi Internet access is becoming popular in many parts of the world by groups of cooperative individuals [17]. There is also a provision of using outdoor APs with higher gain antennae to extend the coverage, but still limit the maximum effective isotropic radiated power (EIRP) within the legal limit. The normal Wi-Fi access point (802.11b or 802.11g) covers only 300 ft, which is roughly 0.0102 square mile; however, the outdoor coverage can be increased by using higher gain antennae [20]. It is also feasible to extend the coverage up to a kilometer, by bearing additional cost on smart antennae. The value of the neighborhood area network is proportional to the number of users in the neighborhood area. By sharing the last-mile infrastructure with multiple users inside the neighborhood coverage area, the network could maximize the return on the investment, thereby leverage the infrastructure investment and fixed cost across the maximum number of neighborhood users.

Campus area network: Campus level networks could include infrastructure that focuses on supporting university, school, organizational campuses, shopping complexes, etc. In this type of network, the majority of buildings are multidwelling units or apartment complexes. WiMAX can coexist with Wi-Fi to deliver megabits of data to the campus area. Following that, Wi-Fi can be used to distribute services to the individual shops, halls, office rooms, lobbies, conference rooms, etc. within the building. Though the WiMAX standard does not describe how much capacity an operator can feed each access point, a single WiMAX BS could handle hundreds of megabits per second of data and can feed one or more Wi-Fi APs mounted on tall buildings inside the campus area. The value of a particular campus area Wi-Fi network depends on the number of users inside the building.

Mini hi-tech parks for suburban cities: Hi-tech habitats would be built not only in prominent cities but also in rural and up-and-coming suburban locations, if proper connectivity is provided. India's IT-export activities depend mainly on the infrastructure erected in urban cities. As a result, those cities are expected to deal with congestion in departments ranging from power to housing. Since metro cities are at a saturation point, the future growth in the IT off-shoring industries will have to come from entirely new towns outside of tier I metro cities.* To promote the growth of IT all over the country, it is imperative that hi-tech campuses are built in and around all major towns. Such mini-IT campus parks would be extremely useful to promote the growth of IT-enabled services—a sector that could provide jobs to millions of Indians. The focus on broadband infrastructure in tier II cities to meet the needs of small IT parks and BPO centers will provide a wider location and enable geographical expansions for IT industries. Each small town can have multiple small hi-tech campus parks. This small hi-tech campus will have the same kind of infrastructure as that of the campus area network discussed above.

Village area network: The cost of a village network will be determined by the choice of the backhaul, since fiber-based and VSAT-based networks have different cost structures. The main goal is to tap India's relatively well-laid fiber infrastructure that penetrates most towns and brings low-cost connectivity to surrounding villages, some of which border the suburban area. A terrestrial fiber backbone always costs less compared to VSAT network. Therefore, villages near a town can take advantage of the fiber backbone; a remote village can be connected via a VSAT link. From the fiber backbone, a point-to-point (PTP) or PMP WiMAX link can be used to connect one or more villages near the town, thus, enabling it to distribute locally among all rural community groups in a given village using long distance Wi-Fi technology.

8.4   Applications of Countrywide Wi-Fi/WiMAX Deployment

A network infrastructure with pure connectivity alone is not enough to enhance the socio-economic class of any given community. Therefore, a simultaneous development of innovative applications and new service models are needed. As ubiquitous wireless technologies and services continue to expand, it is necessary to design new and appropriate applications. The social goal of ubiquitous connectivity is to provide increased access to information for all classes of the community, while its economic goal is to develop information as a commodity along with knowledge products and services. The confluence of these two goals brings together people, information infrastructure, content, and applications.

Applications driving ubiquitous connectivity in metropolitan areas: The quality of urban infrastructure in India, even in the metropolitan cities, is not sophisticated enough to compete with the global economic activity. Though metropolitan cities contribute more to the emerging economy, to accelerate growth, an advanced ICT infrastructure backed by ubiquitous broadband connectivity is necessary. Metro-zone wireless access can become a crucial

---

* http://www.mckinsey.com/ideas/articles/Nasscom_3_Executive_summary.pdf

aspect of a successful strategy to grow the tourism industry. This infrastructure brings low-cost Internet services not only to local residents but also to temporary visitors. Indian cities (e.g., Bangalore, Chennai, Hyderabad, Mumbai, Delhi) have been getting worldwide attention because of their Global IT services and businesses. The transportation infrastructure is so poor in these metro areas that an affordable citywide wireless broadband access could be used as an incentive to encourage telecommuters. Telecommuters will have the convenience to conduct their business from home and can adopt broadband as a necessary tool. Customers can also choose from various access speed options ranging from a guaranteed 128 Kbps to several Mbps, depending on their needs. This infrastructure would be especially helpful in urban settings, as it would be more productive for some employees to work from their home or elsewhere, thus enabling them to save on commuting cost and time. This will aid in promoting the local economy by providing easy Internet access to businesses and especially to the mobile work force. The metro-zone broadband wireless network has more potential for small businesses than its counterpart cable or DSL. The last-mile Wi-Fi network in urban settings will allow local businesses with tools to manage and sell their products and services more effectively. Broadband helps businesses to save on telephone cost through the use of broadband phones (i.e., VoIP); they can also use videoconferencing to save on travel expenses [12]. With modern communication infrastructure, business operations could move away from the main cities to less dense suburban areas, as people can work from any location virtually, eliminating the need for face-to face contact [13].

Applications and service model innovation for suburban economy: Most of the hi-tech parks and the IT service economy are focused toward the urban segment of people, thereby neglecting growth in the suburban, or, the so-called tier II cities. To have unified growth and to make small towns and cities a part of growing knowledge economy, there needs to be some form of ICT infrastructure that can address all developmental issues. A small town can have a mini-IT park that would not need gigabytes of broadband connectivity, but connectivity in megabytes would suffice. This can be easily accomplished with the support of WiMAX technology. Another application of this infrastructure would be neighborhood-based networks. Since many small businesses and activities are simply neighborhood-based, there is a need to localize the network and content so that it appeals to the local community. There can also be unique services and content for each neighborhood. This type of network also hopes to serve as a forum for idea exchange, education, and community enrichment [16]. Each neighborhood having access to their network can form an association using their own infrastructure to develop their communities.

Applications and e-eradication driving rural wireless deployment: The infrastructure for rural environments should have multifunctional communication capabilities and at the same time, should be robust and sustainable. Generally, multipurpose community Internet kiosks are preferred instead of individual household connections. The kiosks can be used to, among other things, send bill payment to various government departments, file complaints, and follow up on electronic applications. Shared hot spots are also the best model where people gather in a particular place in a village for various activities. ICT provides effective tools and techniques for a wide variety of applications such as e-health, e-learning, e-governance, e-entertainment, etc.

e-Education: Affordable broadband technologies such as WiMAX and Wi-Fi will create the possibility of new ways of teaching and learning for rural citizens. Wireless technologies can connect rural schools and colleges to urban institutes, spreading education to the rural areas in a much more pervasive manner than before. This will also enabling the betterment of the education system by eliminating the paucity of teachers in remote areas.

e-Health: Connectivity for health facilities has been identified as a priority to enhance the quality of health care in many ICT projects around the world.* An interesting health application of wireless technologies would be the linking of a rural clinic to a bigger hospital. This would enable data, voice, and video transmission between a rural patient and the city doctor.

Farmers connecting to commodity markets: Farmers in India have been perennially affected by the fluctuations of the commodity market [18]. The information needed to manage risks, track price updates, and prevalent trends in commodity trading in the volatile global market is not available to them. Connectivity will help them to check the weather forecast and register prices of their agro-products at the nearest government market or the futures exchange.† Farmers can also purchase fertilizers, herbicides, and other raw materials needed for their agricultural practice.

e-Village: A village portal enables selling local handicrafts, agriculture/horticulture, and other local products through the website. This portal can also provide comprehensive information about a particular village.

8.5 Conclusion

The growth of emerging countries, like India, in the coming years will be increasingly driven by knowledge- and service-based sectors. The ease of information flow will be a key determinant for success. Deployment and implementation of an affordable communication infrastructure with emerging wireless technologies could be the first step toward narrowing the digital divide. To provide the best connectivity in a very short period, emerging wireless technologies must be well positioned to reach all the villages, towns, and cities of India, thereby enabling a modern hi-tech network infrastructure across the country. This kind of fully integrated modern broadband wireless infrastructure throughout all tiers of the economy will foster equal and sustainable socioeconomic development. We strongly believe that ICT backed by

---

* www.cda.co.za
† http://www.echoupal.com/echoupalwebapplication/

modern wireless technologies will take any developing country into a new age of information economy and wealth creation. Finally, we believe that the national policymakers' core objective should be to develop suitable strategies to promote Wi-Fi and WiMAX technologies giving ubiquitous connectivity.

Acknowledgments

Many thanks to Dr. A. Curtis, Director for Telecommunications and Project Management, Dr. K. Ryan, Professor of Telecommunications Management, and Dr. E. A. Friedman, Director of the Center for Technology Management for Global Development, all at Stevens Institute of Technology, for their valuable input regarding wireless technologies and its applications in developing countries. We would also like to thank Dr. N. K. Shankaranarayan and Dr. Byoung Jo J. Kim, both with AT&T Labs, NJ, USA, for their valuable comments in understanding the economics and technical aspects of infrastructure mesh topologies.

References

1. J. Bjorkdahl, E. Bohlin, and S. Lindmark, Financial assessment of fourth generation mobile technologies, *Communication and Strategies*, no. 54, 2nd quarter, pp. 71–94, 2004.
2. E. Brewer, M. Demmer, B. Du, M. Ho, M. Kam, S. Nedevschi, J. Pal, R. Patra, S. Surana, and K. Fall, The case for technology in developing regions, *IEEE Computer Society*, vol. 38, no. 6, pp. 25–38, 2005.
3. G. Camponova, M. Heitmann, K. S. Slabeva, and Y. Pigneur, Exploring the WISP industry swiss case study. *Presented in 16th Bled Electronic Commerce e-transformation*, Bled, Slovenia, June 9–11, 2003.
4. K. Chebrolu, B. Raman, and S. Sen, Long-distance 802.11b links: Performance Measurement and Experience, in *Proceedings of the 12th annual international conference on mobile computing and networking MobiCom'06*, pp. 74–85, 2006.
5. I. P. Chochliouros and A. S. Spiliopoulou–Chochliourou, Broadband access in European union: An enabler for technical progress, business renewal & social development, *International Journal for Infonomics*, vol. 1, no. 1, pp. 1–21, 2005.
6. C. Corderio, H. Gossain, R. Ashok, and D. Agarwal, The last mile: Wireless technologies for broadband and home networks, in *Proceedings of 21st Brazilian Symposium on Computer Networks*, pp. 19–23, 2003.
7. C. Eklund, R. Marks, K. Stanwood, and S. Wand, IEEE standard 802.16: A technical overview of the wireless MAN air interface for broadband wireless access, *IEEE Communication Magazine*, vol. 40, no. 6, pp. 98–107, 2002.
8. A. M. Elvidge and J. Martucci, Telecommunications network total cost of ownership and return on investment modeling, *BT Technology Journal*, vol. 21, no. 2, pp. 184–190, 2003.

9. E. Evans, J. Lebkowsky, L. Welter, G. Huang, D. Mayfield, and H. Gangadharbatla, *Austin Wireless Future*, IC² Institute, University of Texas, Austin, 2004.
10. V. Gunasekaran and F. Harmantzis, Financial assessment of city wide Wi-Fi deployment, *Communications and Strategies*, no. 63, 3rd quarter, pp. 131–153, 2006.
11. V. Gunasekaran and F. Harmantzis, Emerging wireless technologies for developing countries, *Journal of Technology in Society*, vol. 29, no. 1, pp. 23–42, 2007.
12. B. Lai and G. A. Brewer, New York City's broadband problem and the role of municipal government in promoting a private-sector solution, *Technology in Society*, vol. 28, nos. 1–2, pp. 245–259, 2006.
13. M. L. Moss, S. M. Kaufman, and A. M. Townsend, The relationship of sustainability to telecommunications, *Technology in Society*, vol. 28, nos. 1–2, pp. 235–244, 2006.
14. A. Pentland, R. Fletcher, and A. Hasson, DakNet: Rethinking connectivity in developing nations, *IEEE Computer Society*, vol. 37, no. 1, pp. 78–83, 2004.
15. C. K. Prahalad, *The Fortune at the Bottom of the Pyramid*, Wharton School Publishing, NJ, USA, 2004.
16. B. Rao and M. A. Parikh, Wireless broadband drivers and their social implications, *Technology in Society*, vol. 25, no. 25, pp. 477–489, 2003.
17. C. Sandvig, An initial assessment of cooperative action in Wi-Fi networking, *Telecommunications Policy*, vol. 28, nos. 7–8, pp. 579–602, 2004.
18. D. M. Upton and V. A. Fuller, *The ICT e-Choupal Initiative*, Harvard Business School, case N9-604-016, 2004.
19. K. Wanichkorn and M. Sirbu, The role of fixed wireless access networks in the deployment of broadband services and competition in local telecommunication markets, *Presented at Telecommunications Policy Research Conference (TPRC)*, Alexandria, VA, 28–29, Sept. 2002.
20. M. Zhang and R. Wolff, Using Wi-Fi for cost-effective broadband wireless access in rural and remote areas, in *Proceedings of Wireless Communications and Networking Conference, IEEE WCNC 2004*, vol. 3, pp. 1347–1352, 2004.

9

Connectivity and Load Distribution in WiMAX-Based Multihop Backhaul Networks

Sayandev Mukherjee and Dan Avidor

CONTENTS

9.1   Introduction

We study a large wireless network involving three hierarchical layers. At the lowest layer we have subscriber stations, called "nodes" in the sequel. Each node contains a transceiver with certain common wireless capabilities. Nodes are typically installed at customer premises, and their locations cannot be predicted at the planning stage. The second hierarchical layer contains access points (APs) owned and installed by the service provider wherever he sees fit subject to such constraints as space availability, the parameters of the service provided (e.g., peak data rate), cost considerations, and the expected local penetration of the service. Like regular nodes, APs have no connection to the wired infrastructure and are all identical. To reach the wired infrastructure, e.g., the Internet, the APs rely on high data rate links to gateways (GWs), which form the third hierarchical layer. The GWs are connected to a fiber-optic infrastructure that, for all practical purposes, can be seen as having infinite capacity. Each GW is equipped with high antenna tower, requires significant real estate, and investment. Since GWs are co-located with fibers, and for economic reasons are expected to be relatively scarce, their locations cannot be positioned on the basis of wireless coverage considerations only. To account for this reality, we assume that the GWs, as well as the APs and nodes, are also placed uniformly and randomly over the service area. This location model should not be seen as a worst case scenario, but it is common in the literature and facilitates analysis. The density of APs is expected to be significantly lower than the density of nodes, and the density of GWs even lower.

As said above, each node receives its service through one, or a succession of APs, and one GW. It is highly desirable that a node purchased and installed by a customer will be able to choose an AP based only on measurements of pilot signals emitted by the area's APs. Forcing a node to connect to an AP through an inferior link based on load balancing considerations, for instance, is clearly undesired. The two upper layers of the network containing only APs and GWs may, on the contrary, be designed by the service provider who owns the network, so as to optimize performance. This part of the network must be functional at the start of commercialization, when the final load and distribution of customers is still unknown. Clearly, the statistical model described above cannot predict the situation in any one locality, but it provides insight as to what may happen when the spatial density of registered subscribers grows.

In this chapter, we focus on the basic tree-like (in fact, a forest) backhaul architecture, where a GW constitutes the root of each tree (see Figure 9.1). We investigate the distribution of the APs' load and characterize the required capacity of AP-to-AP and AP-to-GW links to limit the occurrence of overload conditions, which might be expensive to fix. Note that in our model, a node cannot communicate directly with a GW because of hardware incompatibility.

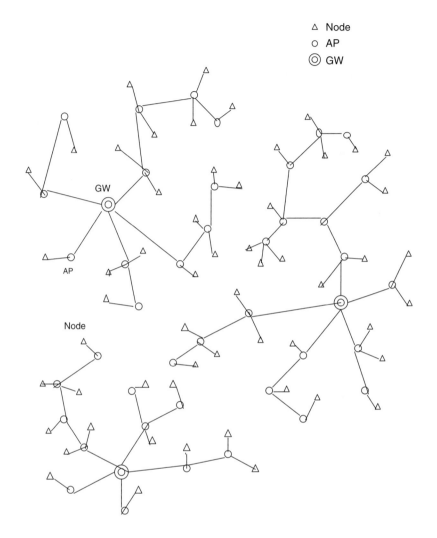

FIGURE 9.1
The "tree-like" backhaul architecture with each tree rooted at a GW.

9.2 Review of Prior Work

Many prior studies focused on the issue of selecting an optimal topology for the backhaul network. Since in such networks a small number of links carry heavy traffic, it is important to achieve uninterrupted service and excellent survivability similar to those achievable with wired infrastructure. Issues such as components failures, irregular congestion conditions, and abnormal propagation conditions must be taken into account, while at the same time the cost must be controlled [1–3]. These contributions consider the tree structure, and also the ring and the multiring topologies, which are common in optical systems. In Ref. 1, the authors propose a self-healing ring topology by providing two disjoint paths between nodes on the ring. In Ref. 2, a mesh-based restorable network topology with link restoration is proposed. The authors formulate the network design problem as an integer programming problem, and also provide an efficient heuristic algorithm, which is more practical for large networks. Another heuristic topology enhancement method that adds redundant spans and upgrades existing infrastructure cost-effectively is proposed in Ref. 3.

9.3 Device Location Model

We begin with the following model assumptions:

1. Nodes are points of a homogeneous Poisson process on the plane with intensity λ_n, i.e.,
   a. The number of nodes in any finite region A, $N_n(A)$, is a Poisson random variable with mean given by $\lambda_n \times \text{area}(A)$, i.e.,

$$\mathbb{P}\{N_n(A) = n\} = e^{-\lambda_n \times \text{area}(A)} \frac{[\lambda_n \times \text{area}(A)]^n}{n!}, \quad n = 0, 1, \ldots$$

(9.1)

   b. Conditioned on a given number of nodes in a chosen region, say $N_n(A) = n$, the locations $(X_1, Y_1), \ldots, (X_n, Y_n)$ of these n nodes are independent and identically distributed (i.i.d.) uniformly over that region with common probability density function (pdf):

$$f_{X_1, Y_1}(x_1, y_1) = \begin{cases} 1/\text{area}(A), & (x_1, y_1) \in A \\ 0, & \text{otherwise} \end{cases}$$

(9.2)

   c. The numbers of nodes in any two disjoint finite regions A and B are independent, i.e.,

$$A \cap B = \emptyset \Rightarrow (\forall m)(\forall n)$$
$$\mathbb{P}\{N_n(A) = m, N_n(B) = n\}$$
$$= \mathbb{P}\{N_n(A) = m\} \times \mathbb{P}\{N_n(B) = n\}$$

(9.3)

2. APs are points of a homogeneous Poisson process on the plane with intensity λ_{AP}, i.e., Equations 9.1 through 9.3 hold with $N_{AP}(\cdot)$ and λ_{AP} in place of $N_n(\cdot)$ and λ_n, respectively.

3. GWs are points of a homogeneous Poisson process on the plane with intensity λ_{GW}, i.e., Equations 9.1 through 9.3 hold with $N_{GW}(\cdot)$ and λ_{GW} in place of $N_n(\cdot)$ and λ_n, respectively.

4. The node, AP, and GW processes are all independent: for any three regions A, B, and C, and any k, l, m, we have

$$\mathbb{P}\{N_n(A) = k, N_{AP}(B) = l, N_{GW}(C) = m\}$$
$$= \mathbb{P}\{N_n(A) = k\}\,\mathbb{P}\{N_{AP}(B) = l\}\,\mathbb{P}\{N_{GW}(C) = m\}$$

9.4 Radio Propagation Model

We assume the following:

1. Attenuation with distance and log-normal shadow fading.

2. The shadow fade attenuation on any link is the same if the transmitter and receiver on that link are interchanged. Since all shadow fades are assumed log-normal, the shadow fade on any link (in either direction) may be represented by $10^{Z/10}$, where Z is a Gaussian random variable with mean zero and variance σ^2, denoted $Z \sim \mathcal{N}(0, \sigma^2)$.

3. The shadow fade attenuations between any node and AP, between any two APs, or between any AP and GW, are all i.i.d. log-normal.*

4. All nodes are identical, all APs are identical, all GWs are identical, and their maximum transmit powers are $P^n_{T,max}$, $P^{AP}_{T,max}$, and $P^{GW}_{T,max}$, respectively.

5. A node i has a connection to an AP b if and only if the received power at the AP on the uplink exceeds some given threshold P^{AP}_{min} and the received power at the node on the downlink exceeds P^n_{min}, i.e., if and only if

$$K^{AP}_n \frac{P^n_{T,max}}{r^\delta_{ib}} 10^{z_{ib}/10} > P^{AP}_{min} \quad \text{and} \quad K^{AP}_n \frac{P^{AP}_{T,max}}{r^\delta_{ib}} 10^{z_{ib}/10} > P^n_{min}$$

$$\Leftrightarrow r_{ib} < r_n 10^{z_{ib}/(10\delta)} \tag{9.4}$$

$$r_n \equiv \left[K^{AP}_n \min\left(\frac{P^n_{T,max}}{P^{AP}_{min}}, \frac{P^{AP}_{T,max}}{P^n_{min}} \right) \right]^{1/\delta} \tag{9.5}$$

---

* The i.i.d assumption is widely used in the literature, even though field measurements seem to indicate that the shadow fades between two links with a common node are correlated [4–7]. However, the authors are not aware of any correlation model that has been widely endorsed by the scientific community.

where δ is the distance-loss exponent, $10^{z_{ib}/10}$ the shadow fade between AP b and node i, r_{ib} the distance between AP b and node i, and K_n^{AP} a constant taking into account the parameters of a node-to-AP link, such as the antenna gains and heights of both the AP and the node. Observe that Equation 9.4 reduces to $r_{ib} < r_n$ if there is no shadow fading, i.e., if $z_{ib} \equiv 0$.

6. AP i has a connection to (i.e., is one hop away from) another AP j if and only if the received power exceeds some given threshold $P_{\min}^{AP}$, i.e., if and only if

$$K_{AP}^{AP} \frac{P_{T,\max}^{AP}}{r_{ij}^{\delta}} 10^{z_{ij}/10} > P_{\min}^{AP} \Leftrightarrow r_{ij} < r_{AP}10^{z_{ij}/(10\delta)},$$

$$r_{AP} \equiv \left(\frac{P_{T,\max}^{AP} K_{AP}^{AP}}{P_{\min}^{AP}} \right)^{1/\delta} \tag{9.6}$$

where $10^{z_{ij}/10}$ is the shadow fade between APs i and j, r_{ij} is the distance between the APs, and K_{AP}^{AP} is defined similarly to K_n^{AP}. Observe that Equation 9.6 reduces to $r_{ij} < r_{AP}$ if there is no shadow fading, i.e., if $z_{ij} \equiv 0$.

7. An AP i has a connection to a GW g if and only if the received power at the GW on the uplink exceeds some given threshold $P_{\min}^{GW}$ and the received power at the AP on the downlink exceeds $P_{\min}^{AP}$, i.e., if and only if

$$K_{AP}^{GW} \frac{P_{T,\max}^{AP}}{r_{ig}^{\delta}} 10^{z_{ig}/10} > P_{\min}^{GW} \quad \text{and} \quad K_{AP}^{GW} \frac{P_{T,\max}^{GW}}{r_{ig}^{\delta}} 10^{z_{ig}/10} > P_{\min}^{AP}$$

$$\Leftrightarrow r_{ig} < r_{GW}10^{z_{ig}/(10\delta)} \tag{9.7}$$

$$r_{GW} \equiv \left[K_{AP}^{GW} \min \left(\frac{P_{T,\max}^{AP}}{P_{\min}^{GW}}, \frac{P_{T,\max}^{GW}}{P_{\min}^{AP}} \right) \right]^{1/\delta} \tag{9.8}$$

where $10^{z_{ig}/10}$ is the shadow fade between the AP and the GW, r_{ig} is the distance between the AP and the GW, and K_{AP}^{GW} is defined similarly to K_n^{AP}. Observe that Equation 9.7 reduces to $r_{ig} < r_{GW}$ if there is no shadow fading, i.e., if $z_{ig} \equiv 0$.

9.5 Connectivity between Network Components

9.5.1 Mathematical Preliminaries

Several calculations in this chapter rely on the following well-known result from the theory of Poisson random variables.

THEOREM 9.1 [19]
Let the number of objects N in a given region be a Poisson random variable with mean μ. Suppose that for any given number of these objects in the region, each of these objects has the same probability p, independent of all the other objects, of having a desired property. Then the number of objects (out of these N) that have the desired property is a Poisson random variable with mean μp.

9.5.2 Distribution of the Number of Neighbors of a Given Node/AP/GW

Define the disk with radius r centered at (x, y) by

$$B(x, y; r) = \{(x', y') : (x' - x)^2 + (y' - y)^2 \leq r^2\}$$

and the punctured disk obtained by deleting the center:

$$B'(x, y; r) = B(x, y; r) \smallsetminus \{(x, y)\}$$

For brevity, we shall write $B(r)$ and $B'(r)$ in place of $B(0,0;r)$ and $B'(0,0;r)$, respectively.

Consider a node at $(0,0)$, and let $M'(r_0)$ be the number of APs whose distance from this node is at most r_0 (where r_0 is assumed to be very large). Since the AP locations are modeled by a homogeneous Poisson point process, we have

$$M'(r_0) \sim \text{Poisson}(\lambda_{AP} \pi r_0^2)$$

Let R denote the distance of an arbitrary AP in $B'(r_0)$ from the node at $(0,0)$. Since the AP is in $B'(r_0)$, from the theory of homogeneous Poisson point processes it follows that its location must be uniformly distributed in $B'(r_0)$, and independent of the location of any other AP(s) in $B'(r_0)$. Thus the pdf of R is

$$f_R(r) = \begin{cases} \dfrac{2r}{r_0^2}, & 0 < r \leq r_0 \\ 0, & r > r_0 \end{cases} \tag{9.9}$$

Let $10^{Z/10}$ be the shadow fade attenuation on the link between an arbitrary AP in $B'(r_0)$ and the node at $(0,0)$. Since the shadow fades are assumed i.i.d. across links, and the locations of the APs in $B'(r_0)$ are also i.i.d., it follows that the event that there is a connection between the arbitrary AP and the node at $(0,0)$ is also i.i.d. across APs, with the common probability given by

$$p_n^{AP}(r_0) = \mathbb{P}\left\{R < r_n 10^{Z/(10\delta)}\right\}$$

$$= \mathbb{P}\left\{R < r_n \exp\left(\frac{hZ}{\delta}\right)\right\} \quad \left[\text{where } h = \frac{\ln 10}{10}\right] \tag{9.10}$$

$$= \mathbb{E}_Z\left[\int_0^{\min\{r_0, r_n \exp(hZ/\delta)\}} \frac{2r}{r_0^2}\, dr\right]$$

$$= \frac{1}{r_0^2}\mathbb{E}_Z\left[\min\left\{r_0^2, r_n^2 \exp\left(\frac{2hZ}{\delta}\right)\right\}\right] \tag{9.11}$$

It then follows from Theorem 9.1 that $M(r_0)$, the number of APs in $B'(r_0)$ with a connection to the node at $(0,0)$, has the following Poisson distribution:

$$M(r_0) \sim \text{Poisson}\left(\lambda_{AP}\pi r_0^2 p_n^{AP}(r_0)\right) \tag{9.12}$$

Now, observe that as $r_0 \to \infty$, we have

$$\lim_{r_0 \to \infty} r_0^2 p_n^{AP}(r_0) = \lim_{r_0 \to \infty} \mathbb{E}_Z\left[\min\left\{r_0^2, r_n^2 \exp\left(\frac{2hZ}{\delta}\right)\right\}\right]$$

$$= \mathbb{E}_Z\left[r_n^2 \exp\left(\frac{2hZ}{\delta}\right)\right]$$

$$= r_n^2 \exp\left(\frac{2h^2\sigma^2}{\delta^2}\right) \tag{9.13}$$

where we use the fact that $Z \sim \mathcal{N}(0, \sigma^2)$ and the known moment generating function (mgf) of the Gaussian distribution.

From Equations 9.13 and 9.12, we see that M, the number of APs in the entire plane with a connection to the node at $(0,0)$, has the following distribution:

$$M \sim \text{Poisson}\left(\mu_n^{AP} \exp(2\alpha^2)\right) \tag{9.14}$$

where

$$\mu_n^{AP} = \lambda_{AP}\pi r_n^2 \tag{9.15}$$

$$\alpha = \frac{h\sigma}{\delta} \tag{9.16}$$

Note that μ_n^{AP} is the mean number of connected neighbor APs of a node in the absence of fading ($\sigma = 0$). Observe that if we redo the above analysis interchanging node and AP we obtain the result that K, the number of nodes in the entire plane with a connection to a given AP, has the following distribution:

$$K \sim \text{Poisson}\left(\mu_n \exp(2\alpha^2)\right) \tag{9.17}$$

where

$$\mu_n \equiv \lambda_n \pi r_n^2 \tag{9.18}$$

From Equation 9.14, it follows that the probability that a node at $(0,0)$ has no connection to any of the APs in the network is given by

$$q_n^{AP} \equiv \mathbb{P}\{M = 0\} = \exp\left[-\mu_n^{AP} \exp(2\alpha^2)\right] \tag{9.19}$$

If we consider an AP instead of a node at $(0,0)$, then the same analysis as above shows that N_{AP}, the number of APs that are connected to the given AP in one hop, has the following distribution:

$$N_{AP} \sim \text{Poisson}\left(\mu_{AP} \exp(2\alpha^2)\right) \tag{9.20}$$

where

$$\mu_{AP} = \lambda_{AP} \pi r_{AP}^2 \tag{9.21}$$

is the mean number of one-hop neighbor APs of an AP in the absence of fading. Further, the probability that an AP at $(0,0)$ has no direct connection to any other AP is given by

$$q_{AP} = \mathbb{P}\{N_{AP} = 0\} = \exp\left[-\mu_{AP} \exp(2\alpha^2)\right] \tag{9.22}$$

In Figure 9.2, we plot q_n^{AP} and q_{AP} versus λ_{AP} for an example where $r_n = 1$, $r_{AP} = 2r_n$, $\delta = 4$, and $\sigma = 8$ dB.

Similarly, N_{GW}, the number of GWs to which an AP at $(0,0)$ has a connection, has the following distribution:

$$N_{GW} \sim \text{Poisson}\left(\mu_{GW} \exp(2\alpha^2)\right) \tag{9.23}$$

where

$$\mu_{GW} = \lambda_{GW} \pi r_{GW}^2 \tag{9.24}$$

is the mean number of one-hop neighbor GWs of an AP in the absence of fading, and the probability that an AP at $(0, 0)$ has no direct connection to any GW is given by

$$q_{GW} = \mathbb{P}\{N_{GW} = 0\} = \exp\left[-\mu_{GW} \exp(2\alpha^2)\right] \tag{9.25}$$

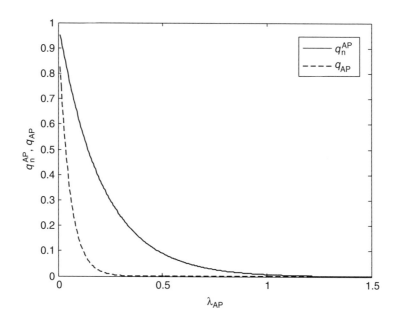

FIGURE 9.2
Plots of q_n^{AP} and q_{AP} versus λ_{AP} for an example where $r_n = 1$, $r_{AP} = 2r_n$, $\delta = 4$, and $\sigma = 8$ dB.

9.5.3 Distribution of the Number of Nodes Supported by a Single AP

Here we focus not on the number of nodes to which a given AP has a connection, but on the actual number of nodes that use their links to this AP to communicate with GWs.

Every node chooses an AP to serve as its connection to a GW on the basis of link quality, or length of the link to the AP, for example. Whatever the basis, we assume that the probability that a certain AP is chosen is i.i.d. across the APs the node has a connection to. The chosen AP is said to *support* the given node. Suppose an arbitrary AP has connections to K nodes, labeled $1, 2, \ldots, K$. Given that node k has a connection to M_k APs, the probability that it chooses this AP is $1/M_k$. We want to find the probability that N nodes ($0 \leq N \leq K$) will choose this AP to support them. Considering each distinct set of N such nodes as an event, there are $\binom{K}{N}$ mutually exclusive events. The probability that $N = n$ is therefore

$$\mathbb{P}\{N = n | K, M_1, \ldots, M_K\} = \sum_{j=1}^{\binom{K}{n}} \prod_{i=1}^{n} \frac{1}{M_i} \prod_{l=n+1}^{K} \left(1 - \frac{1}{M_l}\right)$$

We assume that the event that a node chooses one particular AP out of the APs it has a connection to is independent of the choices of other nodes. Also for any $K, M_1, \ldots, M_K$ are i.i.d. and their common distribution does not depend on K.

Let M be a random variable with this common distribution. As in the previous section, M is the number of APs that an arbitrary node has a connection to, but here its distribution is different, because the node in question is known to have a connection to the given AP, i.e., $M \geq 1$. Hence:

$$\mathbb{P}\{M = m\} = \frac{\exp\left[-\mu_n^{AP}\exp(2\alpha^2)\right]}{1 - \exp\left[-\mu_n^{AP}\exp(2\alpha^2)\right]}\frac{\left[\mu_n^{AP}\exp(2\alpha^2)\right]^m}{m!}, \quad m = 1, 2, \ldots$$

(9.26)

Then we obtain

$$\mathbb{P}\{N = n|K\} = \mathbb{E}_{M_1, \ldots, M_K}[\mathbb{P}\{N = n \mid K, M_1, \ldots, M_K\}]$$

$$= \sum_{j=1}^{\binom{K}{n}}\prod_{i=1}^{n}\mathbb{E}\left[\frac{1}{M_i}\right]\prod_{l=n+1}^{K}\left(1 - \mathbb{E}\left[\frac{1}{M_l}\right]\right)$$

$$= \sum_{j=1}^{\binom{K}{n}}\left(\mathbb{E}\left[\frac{1}{M}\right]\right)^n\left(1 - \mathbb{E}\left[\frac{1}{M}\right]\right)^{K-n}$$

$$= \binom{K}{n}\left(\mathbb{E}\left[\frac{1}{M}\right]\right)^n\left(1 - \mathbb{E}\left[\frac{1}{M}\right]\right)^{K-n}$$

$$\sim \text{Binomial}\left(K, \mathbb{E}\left[\frac{1}{M}\right]\right)$$

(9.27)

Now recall from Equation 9.17 that

$$K \sim \text{Poisson}(\mu_n\exp(2\alpha^2))$$

Thus, the unconditional pmf of N is given by

$$\mathbb{P}\{N = n\} = \mathbb{E}_K[\mathbb{P}\{N = n \mid K\}]$$

$$= \sum_{k=1}^{\infty}\mathbb{P}\{K = k\}\binom{k}{n}\left(\mathbb{E}\left[\frac{1}{M}\right]\right)^n\left(1 - \mathbb{E}\left[\frac{1}{M}\right]\right)^{k-n}$$

$$= \exp\left\{-\mu_n\exp(2\alpha^2)\mathbb{E}\left[\frac{1}{M}\right]\right\}\frac{\left\{\mu_n\exp(2\alpha^2)\mathbb{E}\left[\frac{1}{M}\right]\right\}^n}{n!}, \quad n = 0, 1, \ldots$$

$$\sim \text{Poisson}\left(\mu_n\exp(2\alpha^2)\mathbb{E}\left[\frac{1}{M}\right]\right)$$

(9.28)

Note that we could get the same answer directly from Theorem 9.1: the number of nodes that have a connection to the given AP is a Poisson random variable K with mean $\mu_n\exp(2\alpha^2)$, and each of these nodes independently and identically chooses the given AP with probability $\mathbb{E}[1/M]$.

Hence the total number N of nodes that the given AP supports is Poisson with mean $\mu_n \exp(2\alpha^2)\, \mathbb{E}[1/M]$. The pmf of N when the nodes switch On and Off randomly and independently across nodes, with probability p of being in the On state, can also be calculated from Equation 9.28 by replacing $\mu_n \exp(2\alpha^2)\, \mathbb{E}[1/M]$ with $p\mu_n \exp(2\alpha^2)\, \mathbb{E}[1/M]$.

Finally, we evaluate

$$
\mathbb{E}\left[\frac{1}{M}\right] = \sum_{m=1}^{\infty} \mathbb{P}\{M = m\}\frac{1}{m}
$$

$$
= \frac{\exp\left[-\mu_n^{AP}\exp(2\alpha^2)\right]}{1-\exp\left[-\mu_n^{AP}\exp(2\alpha^2)\right]}\sum_{m=1}^{\infty}\frac{\left[\mu_n^{AP}\exp(2\alpha^2)\right]^m}{m\,m!}
$$

$$
= \frac{\exp\left[-\mu_n^{AP}\exp(2\alpha^2)\right]}{1-\exp\left[-\mu_n^{AP}\exp(2\alpha^2)\right]}\left[\mathrm{Ei}\left(\mu_n^{AP}\exp(2\alpha^2)\right)-\gamma-\ln\left(\mu_n^{AP}\exp(2\alpha^2)\right)\right]
$$

$$(9.29)$$

where $\gamma \approx 0.57721$ is the *Euler Gamma constant*,

$$
\mathrm{Ei}(x) = -\int_{-x}^{\infty}\frac{e^{-t}}{t}\,dt
$$

is the *exponential integral*, and we use the known series representation of $\mathrm{Ei}(x)$ in Ref. 10, p. 877.

$$
\mathrm{Ei}(x) = \gamma + \ln x + \sum_{m=1}^{\infty}\frac{x^m}{m\,m!}, \quad x > 0
$$

In Figure 9.3, we plot $\mathbb{E}[1/M]$ versus λ_{AP} for the same example as in Figure 9.2 with $r_n = 1$, $\delta = 4$, and $\sigma = 8\,\mathrm{dB}$.

9.5.4 Probability Distribution of Distance between a GW and an AP with a Direct Connection to the GW

For later reference, we derive here the pdf of the distance between a GW and an AP with a direct connection to the GW. Recall the definition of the indicator function $1_A(\cdot)$ of an arbitrary set A:

$$
1_A(x) = \begin{cases} 1, & \text{if } x \in A \\ 0, & \text{if } x \notin A \end{cases}
$$

For the same situation as analyzed in Section 9.5.2, let us define $p_{AP}^{GW}(r_0)$ to be the probability that an AP at $(0,0)$ has a direct connection to an arbitrary GW uniformly distributed over $B'(r_0)$. Then the pdf of the distance R between

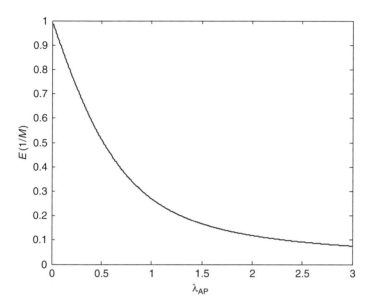

FIGURE 9.3
Plot of $\mathbb{E}[1/M]$ versus λ_{AP} for the same example as in Figure 9.2 with $r_n = 1$, $\delta = 4$, and $\sigma = 8\,\text{dB}$.

the AP and the GW, conditioned on there being a direct connection between them, i.e., $R < r_{GW}\exp(hZ/\delta)$, where $10^{Z/10}$ is the shadow fade on the link between the AP and GW, is given by

$$f_{R \mid R < r_{GW}\exp(hZ/\delta)}(r)$$

$$= \frac{d}{dr}\lim_{r_0 \to \infty} \frac{\mathbb{P}\{R \le \min[r_0, r], R < r_{GW}\exp(hZ/\delta)\}}{p_{AP}^{GW}(r_0)}$$

$$= \frac{d}{dr}\lim_{r_0 \to \infty} \frac{\mathbb{E}_R\left[1_{(0,\min[r_0,r])}(R)Q(\ln(R/r_{GW})/\alpha)\right]}{p_{AP}^{GW}(r_0)}$$

$$= \frac{d}{dr}\lim_{r_0 \to \infty} \frac{1}{r_0^2 p_{AP}^{GW}(r_0)}\int_0^r 2uQ\left(\frac{\ln(u/r_{GW})}{\alpha}\right)du$$

$$= \frac{2rQ(\ln(r/r_{GW})/\alpha)}{r_{GW}^2 \exp(2\alpha^2)} \tag{9.30}$$

where

$$Q(x) = \int_x^\infty \frac{\exp(-u^2/2)}{\sqrt{2\pi}}\,du$$

and we follow the same steps as in the derivation of Equation 9.13.

Similarly, for a given AP, the pdf of the distance R' to any other AP, conditioned on there being a direct connection between them, i.e., $R' < r_{AP} \exp(hZ'/\delta)$, where $10^{Z'/10}$ is the shadow fade on the link between the two APs, is given by

$$f_{R' \mid R' < r_{AP} \exp(hZ'/\delta)}(r') = \frac{2r'Q\left(\dfrac{\ln(r'/r_{AP})}{\alpha}\right)}{r_{AP}^2 \exp(2\alpha^2)} \qquad (9.31)$$

9.6 Measures of Coverage in the Overall Network

Recall that nodes cannot connect to GWs directly, but can do so via one or a succession of APs. We also assume that an AP is supported by a single GW, or a single AP. In contrast to the GW, an AP has a limited bandwidth that cannot be exceeded. The network of APs and GWs is designed by the network operator. An important objective is to lower the probability that parts of the network will have to be replaced as the customer base grows. An attractive topology is obtained when the majority of nodes can reach a GW via only one AP, in other words, in two hops. This is also the most desirable topology in terms of network reliability. We now show how to calculate the probability that an arbitrary node cannot reach a GW in two hops. Note that this is also the fraction of nodes that cannot be reached by a GW in two hops, and which will therefore require the use of more than one AP on the path to a GW.

9.6.1 The Fraction of Nodes That Cannot Be Reached by a GW in Two Hops

The calculation has two stages—first we derive the probability of not being able to connect to a specific GW a distance D away in two hops, then proceed to derive the desired probability by averaging over the distribution of this distance.

9.6.1.1 Probability That a Node Cannot Connect to a Specific GW a Distance D Away in Two Hops

Without loss of generality, let the GW be located at $(D, 0)$. Let us call the probability of interest $q_2(D)$.

Note that there is no connection between the node at $(0, 0)$ and the GW at $(D, 0)$ in two hops if and only if any AP that has a connection to the node at $(0, 0)$ does not have an one-hop connection to the GW at $(D, 0)$. Let us label the shadow fades on the link between an arbitrary AP and the given node at $(0, 0)$ and GW at $(D, 0)$ by Z and Z', respectively. For any given values of Z and Z', this AP has a connection to both the nodes at $(0, 0)$ and the GW at

$(D, 0)$ if and only if it lies in the region of intersection of the disks centered at $(0, 0)$ and $(D, 0)$ with radii $r_n e^{hZ/\delta}$ and $r_{GW} e^{hZ'/\delta}$, respectively. Then we have

$\mathbb{P}\{\text{node at } (0, 0) \text{ cannot communicate with the GW at } (D, 0) \text{ in two hops}\}$

$$= \lim_{r_0 \to \infty} \sum_{k=0}^{\infty} \mathbb{P}\{N_{AP}(B'(0, 0; r_0) \smallsetminus \{(D, 0)\}) = k\}$$

$\times \mathbb{P}\{\text{node at } (0, 0) \text{ cannot communicate with GW at } (D, 0)$

$\text{in two hops} \mid N_{AP}(B'(0, 0; r_0) \smallsetminus \{(D, 0)\}) = k\}$

$$= \lim_{r_0 \to \infty} \sum_{k=0}^{\infty} \exp(-\lambda_{AP}\pi r_0^2) \frac{(\lambda_{AP}\pi r_0^2)^k}{k!}$$

$\times [1 - \mathbb{P}\{\text{An arbitrary AP in } B'(0, 0; r_0) \smallsetminus \{(D, 0)\}$

$\text{has a connection to both the nodes at } (0, 0) \text{ and the GW at } (D, 0)\}]^k$

$$(9.32)$$

where we use the independence of the events that the different APs in $B'(0, 0; r_0) \smallsetminus \{(D, 0)\}$ have connections to both the nodes at $(0, 0)$ and the GW at $(D, 0)$.

Summing the infinite series in Equation 9.32, we can rewrite it as

$\mathbb{P}\{\text{node at } (0, 0) \text{ cannot communicate with a GW at } (D, 0) \text{ in two hops}\}$

$$= \lim_{r_0 \to \infty} \exp(-\lambda_{AP}\pi r_0^2) \exp[\lambda_{AP}\pi r_0^2(1 - \mathbb{P}\{\text{An arbitrary}$$

$\text{AP in } B'(0, 0; r_0) \smallsetminus \{(D, 0)\}$

$\text{has a connection to both the node at } (0, 0) \text{ and the GW at } (D, 0)\})]$

$$= \lim_{r_0 \to \infty} \exp[-\lambda_{AP}\pi r_0^2 \, \mathbb{P}\{\text{An arbitrary AP in } B'(0, 0; r_0) \smallsetminus \{(D, 0)\} \text{ has a}$$

$\text{connection to both the node at } (0, 0) \text{ and the GW at } (D, 0)\}] \qquad (9.33)$

Finally, since this node location is uniformly distributed on the disk $B(0, 0; r_0)$, it follows from Equation 9.4 that

$\mathbb{P}\{\text{An arbitrary AP in } B'(0, 0; r_0) \smallsetminus \{(D, 0)\} \text{ has a}$

$\text{connection to both the node at } (0, 0) \text{ and the GW at } (D, 0)\}$

$$= \frac{\mathbb{E}_{Z,Z'}\left[\text{area}\left(B'(0, 0; r_n e^{hZ/\delta}) \cap B'(D, 0; r_{GW} e^{hZ'/\delta})\right)\right]}{\text{area}(B(0, 0; r_0))}$$

$$= \frac{\mathbb{E}_{Z,Z'}\left[\text{area}\left(B'(0, 0; r_n e^{hZ/\delta}) \cap B'(D, 0; r_{GW} e^{hZ'/\delta})\right)\right]}{\pi r_0^2} \qquad (9.34)$$

Substituting Equation 9.34 into Equation 9.33, we finally obtain

$$
\begin{aligned}
q_2(D) &= \exp\left\{-\lambda_{AP}\mathbb{E}_{Z,Z'}\left[\text{area}\left(B\left(0,0;r_n e^{hZ/\delta}\right)\cap B\left(D,0;r_{GW}e^{hZ'/\delta}\right)\right)\right]\right\} \\
&= \exp\left\{-\frac{\lambda_{AP}}{2\pi}\int_{-\infty}^{\infty}dv\,e^{-v^2/2}\int_{-\infty}^{\infty}dw\,e^{-w^2/2}g\left(D;r_n e^{\alpha w}, r_{AP}e^{\alpha v}\right)\right\}
\end{aligned}
$$

$$(9.35)$$

where $g(r;a,b)$ is the area of intersection of two disks of radii a and b, respectively, whose centers are spaced r apart. From Ref. 11, Equation 11 we have

$$
g(r;a,b) \equiv \text{area}(B(r,0;a)\cap B(0,0;b))
$$

$$
= \begin{cases}
\pi(\min\{a,b\})^2, & r < |a-b| \\[1em]
a^2\cos^{-1}\left(\dfrac{r^2+a^2-b^2}{2ra}\right)+b^2\cos^{-1}\left(\dfrac{r^2+b^2-a^2}{2rb}\right) & \\[0.5em]
\quad -\dfrac{1}{2}\sqrt{[(a+b)^2-r^2][r^2-|a-b|^2]}, & |a-b|\leq r < a+b \\[1em]
0, & r \geq a+b
\end{cases}
$$

$$(9.36)$$

9.6.1.2 Probability That an Arbitrary Node Cannot Connect to Any GW in Two Hops

Consider a node at (0, 0). We know that the number of GWs whose distance from this node is at most r_0 (where r_0 is assumed to be very large) is a Poisson($\lambda_{GW}\pi r_0^2$) random variable. Let R denote the distance of an arbitrary GW in $B'(r_0)$ from the node at $(0,0)$. Then R has pdf

$$
f_R(r) = \begin{cases}
\dfrac{2r}{r_0^2}, & 0 < r \leq r_0 \\[1em]
0, & r > r_0
\end{cases}
$$

$$(9.37)$$

Assume that each of the GWs in $B'(r_0)$ has the same probability $p_2(r_0)$, independent of all the other GWs, of having a connection to the given node in two hops. This may not be strictly correct, because two close-by GWs might obtain a connection in two hops through a common AP that has a single-hop connection to the node at $(0, 0)$. However, when the number of APs is much larger than the number of GWs, it is unlikely that a path between the node and two different GWs will pass through a single AP. We will therefore make this assumption in the sequel.

Note that for an arbitrary GW at distance r from the node at (0, 0), the probability that it has a connection to the given node in two hops is just $1 - q_2(r)$, where $q_2(\cdot)$ is given by Equation 9.35. We assume that r_0 is large enough for us to ignore edge effects (this assumption becomes exact when we let $r_0 \to \infty$). Then we have

$$p_2(r_0) = \int_0^{r_0} [1 - q_2(r)] f_R(r) \, dr = \frac{1}{r_0^2} \int_0^{r_0} 2r[1 - q_2(r)] \, dr \qquad (9.38)$$

An argument similar to that in Section 9.5.2 using Theorem 9.1 shows that M_2, the number of GWs (in the entire plane) with a connection to the node at $(0, 0)$ in two hops, has the following Poisson distribution:

$$M_2 \sim \text{Poisson}(\lambda_{GW}\pi \lim_{r_0 \to \infty} r_0^2 p_2(r_0)) \equiv \text{Poisson}\left(2\pi\lambda_{GW} \int_0^\infty r[1 - q_2(r)] \, dr\right)$$
$$(9.39)$$

Hence the probability that an arbitrary node cannot reach any GW in two hops is given by

$$q = \mathbb{P}\{M_2 = 0\} = \exp\left(-2\pi\lambda_{GW} \int_0^\infty r[1 - q_2(r)] \, dr\right) \qquad (9.40)$$

and this is also the fraction of nodes that will connect to a GW via more than one AP.

Observe from Equations 9.40 and 9.35 that $1 - q$, the probability that an arbitrary node can reach a GW in two hops, can be increased by increasing either λ_{GW} or λ_{AP} or both, but that the effect of these two intensities is different.

It is natural to ask what the probability is that an arbitrary node cannot reach any GW in $\leq t$ hops, say, where $t \geq 3$. Unfortunately, this probability cannot be determined exactly. However, lower bounds on this probability have been derived in Ref. 9.

9.6.2 Design of the AP-to-GW Link to Avoid Overload

9.6.2.1 Distribution of the Traffic Load on an AP Serving as a Root of a Tree Containing Other APs

Some APs support only nodes, and on the uplink connect to another AP or directly to a GW. The rest of the APs support other APs and, possibly, nodes also. Suppose the routing chosen by the service provider is such that an AP supports $m - 1$ other APs, then including itself this is m APs altogether. From Equation 9.28 we know that the number of nodes N_i, $i = 1, \ldots, m$ supported directly by any of these m APs has the following distribution:

$$N_i \sim \text{Poisson}\left(\mu_n \exp(2\alpha^2)\mathbb{E}\left[\frac{1}{M}\right]\right), \quad i = 1, \ldots, m \qquad (9.41)$$

where $\mathbb{E}[1/M]$ is given by Equation 9.29.

Then N, the number of nodes that are supported by the given AP, is given by

$$N = \sum_{i=1}^{m} N_i$$

We assume that the fact that these APs are chained together to form a tree does not affect the distribution of the number of nodes supported by each, as derived in Section 9.5.3. From the above assumption, $N_1, \ldots, N_m$ are all i.i.d. Since this common distribution is Poisson, it follows that N is also Poisson with the following distribution:

$$N \sim \text{Poisson}\left(m\mu_n \exp(2\alpha^2)\mathbb{E}\left[\frac{1}{M}\right]\right) \tag{9.42}$$

Thus if all nodes belong to the same service class and generate the same mean traffic load, then the mean load on this AP is $m\mu_n \exp(2\alpha^2)\mathbb{E}[1/M]$ times the mean traffic load generated by each node, where $\mathbb{E}[1/M]$ is given by Equation 9.29.

9.6.2.2   Overload Condition on an AP

Suppose an AP can support the data rate R_{AP}. Suppose also that $\rho_{AP}R_{AP}$, $0 < \rho_{AP} < 1$ is available for payload data and the rest is required for demultiplexing, overhead, and synchronization tasks. Suppose that the data rate (payload only) of a node-to-AP link is $\rho_n R_n$, $0 < \rho_n < 1$, then

$$L = \left\lceil \frac{\rho_{AP}R_{AP}}{\rho_n R_n} \right\rceil$$

nodes at most can be supported by an AP. Thus the probability of "overload" on this AP is

$$\mathbb{P}\{N > L\} = 1 - F_N(L)$$

where N is distributed as in Equation 9.42.

9.7   Summary

In this chapter, we examine the backhaul requirements of a large fixed wireless network providing high-speed data service to customer premises. We assume a network comprising three hierarchical levels, and a location model common in the literature. Since the backhaul system must be mostly operational before customers can be served, service providers face a trade-off between the desire to minimize the initial investment and the need to dimension the network so as to match the service needs later on as the customer base continues to develop. It is therefore required to estimate the load that every part of the backhaul network will be subjected to. In this chapter, we offer an analytical study of certain aspects of the problem. We start with basic relationships controlling communications between nodes. Defining the service requirements of a single customer as a unit load, we calculate the distribution of the load supported by a single AP, under the assumption that each node chooses an AP based only on the quality of the links to the APs in its neighborhood. We then proceed to calculate the distribution of the load on sections of the backhaul network, and show the effects of the main network parameters: λ_n, λ_{AP}, λ_{GW}, and r_n, r_{AP}, r_{GW}. Since the cost of the various network components could be vastly different, this is an important input to good economical design. In this chapter, we focus exclusively on dimensioning simple tree structured networks, as opposed to other topological structures offering redundant routes and fast restoration capabilities.

References

1. C. Charnsripinyo and N. Wattanapongsakorn, A model for reliable wireless access network topology design, in *Proceedings of the TENCON 2004*, Chiang Mai, Thailand, 2004 IEEE Region 10 Conference Volume B, 21–24, vol. 2, pp. 561–564, Nov. 2004.
2. C. Charnsripinyo and D. Tipper, Topological design of 3G wireless backhaul networks for service assurance, in *Proceedings of the DRCN 2005*, Ischia, Italy, Design of Reliable Communication Networks, pp. 1–9, Oct. 2005.
3. W-S. Soh, Z. Antoniou, and H.S. Kim, Improving restorability in radio networks, in *Proceedings of Globecom 2003*, San Francisco, USA, IEEE Global Telecommunications Conference 2003, vol. 6, pp. 3493–3497, Dec. 2003.
4. J. van Rees. Cochannel measurements for interference limited small-cell planning, *Int. J. Electron. Commun. (Archiv fur Elektronik und Ubertragungstechnik)*, vol. 41, no. 5, pp. 318–320, 1987.
5. V. Graziano, Propagation correlations at 900 MHz, *IEEE Trans. Vehicular Tech.*, vol. 27, no. 4, pp. 182–189, 1978.
6. H.W. Arnold, D.C. Cox, and R.R. Murray, Macroscopic diversity performance measured in the 800-MHz portable radio communications environment, *IEEE Trans. Ant. Prop.*, vol. 36, no. 2, pp. 277–281, 1988.
7. K. Zayana and B. Guisnet, Measurements and modelisation of shadowing cross-correlations between two base stations, in *Proceedings of the* ICUPC, Florence, Italy, 1998, pp. 101–105.
8. J.F.C. Kingman, *Poisson Processes*, Oxford, UK: Oxford University Press, 1993.
9. S. Mukherjee, D. Avidor, and K. Hartman, Connectivity, power and energy in a multihop cellular packet system, *IEEE Trans. Vehicular Tech.*, vol. 56, no. 2, pp. 818–836, Mar. 2007.
10. I.S. Gradshteyn and I.M. Ryzhik, *Table of Integrals, Series, and Products*, 6th ed. San Diego, CA: Academic Press, 2000.
11. E. Weisstein, ed. *Mathworld*. http://mathworld.wolfram.com/Circle-CircleIntersection.html

10

Providing QoS to Real and Interactive Data Applications in WiMAX Mesh Networks

Vinod Sharma and Harish Shetiya

CONTENTS

10.1   Introduction

IEEE 802.16 standard [1], also known as WiMAX has been specifically designed to provide wireless broadband access in the metropolitan area network (MAN), delivering performance comparable to traditional cable, digital subscriber line (DSL), or T1 offerings. To provide the coverage and data rates envisioned, even or uneven terrain, the use of multihop communication seems desirable. Hence, WiMAX supports a mesh mode (the other mode being point-to-multipoint (PMP)) in which unlike the traditional cellular systems, the nodes can communicate without having a direct connection with the base station (BS).

In an IEEE 802.16d mesh network, a node that has a direct connection to backhaul services outside the mesh network is termed a mesh base station (MBS). All other nodes of a mesh network are termed mesh subscriber stations (MSSs). In IEEE 802.16d standards, these nodes are stationary, that is, the standards do not support mobility (see however 802.16e amendment [2] to the 802.16 standard that supports the mobility). The standard specifies a centralized scheduling scheme for mesh networks. Under this scheme, the MSSs notify the MBS their data transfer requirements and the quality of their links to their neighbors. The MBS uses the topology information along with the requirements of each MSS to decide the routing and the scheduling. The multiple access control (MAC) scheme used is time division multiple access (TDMA) and the resource allocation is in terms of time slots within a frame. The standard does not specify an algorithm for scheduling of the slots to different MSSs, neither does it specify any routing algorithm. Scheduling and routing will have significant impact on the performance of the system and will largely decide the end-to-end quality of service (QoS) to different users.

The WiMAX standard also supports a distributed scheduling scheme in which each mesh node uses the local topology, channel, and traffic information to decide which channel to use. The distributed approach is simple and robust as compared to the centralized approach. However, it results in lower channel utilization and will provide less control over QoS. Thus, it is recommended [11] that the distributed scheduling be used only for unlicensed spectrum while the centralized for licensed. The standard recommends the centralized scheduling for traffic entering/leaving the mesh while the distributed scheduling for intranet traffic. A part of the frame can be reserved for centralized scheduling and another for distributed scheduling and these can be configured. Since most of the traffic is expected to be Internet traffic, we will concentrate on centralized mode.

In the following, we survey the literature on scheduling and routing for wireless networks. Scheduling algorithms to provide QoS in single hop (PMP) IEEE 802.16 networks are considered in Refs. 13, 18, 36 and 40. See also Ref. 19 for a performance analysis.

The problem of scheduling and routing in *ad hoc* multihop wireless networks has been extensively studied in recent years (see Refs. 4, 6, 15, 26 and 29 for general surveys and tutorials). Scaling laws for fundamental limits on information transfer in multihop wireless networks are surveyed in Ref. 42. The dominant MAC protocols being considered for multihop networks are the carrier sense multiple access/collision avoidance (CSMA/CA)-based IEEE 802.11 and the TDMA-based IEEE 802.16. Since 802.11 technology is much more mature and cheaper, most of the mesh network deployments are based on it. However, due to co-channel interference it does not provide satisfactory performance [28,41] while 802.16 can be much superior. See Ref. 21 for a recent contribution to provide QoS in 802.11-based mesh networks.

The studies on multihop 802.16 networks are provided in Refs. 8, 9, 23, 34, 35 and 39. In Ref. 39, a simple heuristic scheduling and a tree routing algorithm are proposed to achieve efficient channel utilization. Refs. 8 and 23 provide fair access to all nodes and also efficient utilization of resources. Routing and scheduling algorithms are provided in Ref. 35, which are efficient for the overall system but spatial reuse of the channels is not allowed (because the 802.16 standard at that time did not allow spatial reuse). In Ref. 34 also channel spatial reuse is not allowed but within this limitation they provide QoS to individual transfer control protocol (TCP) and real-time connections. The QoS guarantee to individual flows has not been provided in any other multihop wireless network study that we are aware of (all the other studies mentioned above provide scheduling and routing for the aggregate traffic generated at different nodes, which as we will see is not sufficient to guarantee QoS to individual connections). In Ref. 9, the authors study the distributed scheduling.

In this chapter, we present algorithms for centralized scheduling of real- and nonreal-time traffic with the objective of providing QoS within the framework of the IEEE 802.16 mesh mode. We first obtain an optimal and fair routing and scheduling of the aggregate traffic generated at different nodes within the network. This way we fix the particular real-time and TCP connections that pass through a particular link and also the slots in which the link transmits. Next, we develop algorithms to decide how each link transmits the packets of different flows passing through it on the slots assigned to it so as to provide QoS to individual flows. The real-time applications use user datagram protocol (UDP) while the data applications use TCP. TCP, being window flow controlled, behaves very differently from UDP. First, we develop algorithms for UDP and then for TCP. Finally, we combine these algorithms to provide QoS in a network serving both real and data applications. To ensure sufficient resources we also discuss an admission control algorithm that can be used in our setup. Our algorithms use the network resources efficiently and fairly and can be used in real time by the MBS.

The organization of the chapter is as follows. Section 10.2 describes the system model. We obtain an optimal and fair routing and link scheduling algorithm in Section 10.3. In Section 10.4, we develop scheduling algorithms to provide QoS to UDP connections. TCP connections will be studied in Section 10.5. In Section 10.6, we handle both UDP and TCP traffic together to provide QoS to each connection. Section 10.8 provides an admission control policy. Section 10.9 concludes the chapter.

10.2   System Model

IEEE 802.16 supports two modes of operation: PMP and mesh mode. In PMP, the traffic is transmitted directly between the BS and the subscriber station (SS). This is the common mode and current implementation efforts are directed at PMP. In the mesh mode, the overall area is divided into meshes. Each mesh has an MBS. The other nodes in a mesh are called MSSs. A transmission can take place between two MSSs within a mesh or between two different meshes. The transmission between two MSSs within a mesh can occur via other MSSs within the mesh that may or may not involve the MBS. Transmission between two MSSs in two different meshes involves transmission from the source MSS to its MBS (possibly via other MSSs in the mesh), from MBS to the BS, from the BS to the MBS of the receiver mesh, and finally from this MBS to the receiver MSS.

In this chapter, we consider the mesh mode. We provide a brief overview of the IEEE 802.16 mesh mode of operation and present the system model that we use in our work.

The mesh mode supports two different physical layers, that is, WirelessMAN-orthogonal frequency division multiplexing (OFDM) and WirelessHUMAN. Both of these use 256 point fast fourier transfer (FFT) OFDM TDMA/TDM for channel access and operate in a frequency band below 11 GHz. The first operates in the licensed band but the second uses the unlicensed band. The standards also support adaptive modulation and coding where the burst profile of a link (i.e., modulation scheme and the coding rate) and hence the link rate is changed depending on the channel conditions.

The mesh mode supports only time division duplex (TDD) to share the channel between the uplink and the downlink. A mesh frame consists of a control and a data subframe. The control subframe serves two basic functions. One is the creation and maintenance of cohesion between different stations and the other is the coordinated scheduling of data transfers between stations. The data subframe consists of MAC protocol data units (PDUs) transmitted by different users. A MAC PDU consists of a generic MAC header, a mesh subheader, and optional data. The standards support both centralized and distributed scheduling of slots. Centralized scheduling is mainly used to transfer data between the MBS and the MSSs. Since this is the usual scenario, centralized scheduling is the dominant mode. The MBS periodically collects the channel information and the resource (throughput) requests of all the nodes to draw up the schedule which it distributes to the nodes.

We consider the following scenario. Consider a mesh network with M MSSs labeled $1, 2, \ldots, M$. The MBS is labeled 0. We consider uplink and downlink *Centralized Scheduling* of the MSSs, which, according to the standards, uses TDMA with spectral reuse. Also the data are directed either to or from the MBS. We assume that each node transmits at the maximum allowed power, if needed. (Although power control is also an important issue in performance of a mesh network, the standard currently does not emphasize it and hence we do not address this in this chapter.) As the channel condition on a link changes, the data rate is also changed so as to meet the desired bit error rate (BER). Let r_{ij} denote the rate and $E[r_{ij}]$ the average rate of the channel from node i to node j. Resource allocation is done by the MBS in units of (mini) time slots. One time slot consists of multiple OFDM symbols. Each allocation is valid for K frames consisting of N time slots (for simplicity of notation we will take $K = 1$).

IEEE 802.16 supports real- and nonreal-time applications. The real-time applications, for example, IP telephony and video conferencing use UDP while data applications use TCP. Real-time applications and interactive data (file transfer and web browsing) require QoS. To provide QoS to these applications will require careful routing and scheduling of traffic through the mesh network. We will consider these problems for both types of traffic. Since UDP traffic and real-time QoS requirements are very different from TCP traffic and interactive data QoS requirements, we will consider these problems separately and then show how to integrate them in the same system.

To provide the QoS, we will generally follow the QoS architecture developed in Ref. 34 since this seems to be the only architecture available for 802.16 mesh networks, which guarantees per flow QoS. However, due to 802.16 mesh requirements at that time Shetiya and Sharma [34] did not consider spectral spatial reuse. This can be a serious limitation because spectral reuse can provide significant capacity improvement. Thus, in our current proposal we will remove this restriction.

We will use a two-step approach. In the first step, we will provide routing and scheduling for the aggregate traffic for each source–destination pair of MSSs (one of these MSSs will be the MBS). The aggregate traffic will be the *mean* total traffic of all the real and data applications between different source–destination pairs. This of course does not guarantee the QoS to individual flows. In the second step, we develop scheduling algorithms to share the long-term throughput guaranteed in step one between real and data applications of each source–destination pair to guarantee QoS to individual flows. We will justify the two-step approach and will provide simulation results to verify the claims on QoS guarantees.

Section 10.3 provides the routing and scheduling for step one to satisfy the aggregate demands of source–destination pairs. In Sections 10.4 through 10.6, we detail our step two to ensure the QoS to individual flows.

10.3 Routing and Scheduling for Aggregate Traffic

The algorithms developed in this section can be used for uplink as well as downlink simultaneously. Let $\lambda(s, d)$ be the mean number of bits per slot to be transmitted from MSS s to MSS d. This is the sum of mean throughput required by all the real-time and data connections transmitting from MSS s to MSS d. The calculation of mean throughput requirements for TCP connections is shown in Section 10.5. For the constant bit rate (CBR) connections, it is the traffic they generate per slot. For a variable bit rate (VBR) connection, it equals the equivalent bandwidth (for calculations see Section 10.4.2; h is equal to the maximum distance between the MBS and an SS in the mesh). For downlink MSS s will always be the MBS and for uplink MSS d will be the MBS. We develop algorithms in this section, which will decide the routes that $\lambda(s, d)$ will follow and also the slots in which each link will transmit.

We develop algorithms, which provide routes and schedules that will be functions of $\lambda(s, d)$ and the mean link rates $E[r(i, j)]$ but otherwise would not vary with time. By exploiting the current queue lengths at different links and the channel states one could vary the routes and the schedules to obtain better performance [33,34], but we will not do this. This is because frequently varying routes and schedules in real time are computationally more complex and can also cause loops in the path traversed by a packet and resequencing problem at the receiver. Furthermore, our QoS architecture requires reservation of resources at intermediate MSSs along a route. Thus, frequent route variation is not desirable. In addition, in case of spectral spatial reuse, changing the routes and schedules is more complicated. Thus, we will change the routes and schedules of link transmissions only when some of the $\lambda(s,d)$ or $E[r(i,j)]$ change drastically and some links/nodes fail. These algorithms will be run at the MBS and then the schedules broadcast to different nodes via mesh centralized schedule messages.

The algorithms we develop will satisfy the traffic requirements $\lambda(s,d)$ of each source–destination pair (s,d) if possible. If not, then we will provide a fair solution that is also efficient. When it is possible to satisfy certain (fair) traffic requirements of all source–destination pairs, we will provide routing and scheduling that optimizes a cost function.

In this section, we use the approach developed in Ref. 31, which in turn was partly motivated by Ref. 24.

In Ref. 34, where spatial reuse is not allowed, it was shown that the routing and scheduling problems can be decoupled and that a tree structure can be optimal for routing. In the present general scenario this may not be true (although the 802.16 standard seems to prefer the tree structure [8,39]).

The cost function to optimize will be a sum of the link cost functions $f(\Gamma(i,j)n(i,j))$, where $\Gamma(i,j)$ is the total mean traffic rate per slot and $n(i,j)$ the fraction of slots assigned to link (i,j). Better cost functions can be obtained as a function of higher moments of traffic arriving at link (i,j) but higher moments are difficult to obtain and handle. Thus, it is desirable to use only the first moments. But even then f will often be a nonlinear function. For example, using Kleinrock's independence assumption [38] or approximating the queues at each link by an $M/M/1$ queue, we get

$$f(\Gamma(i,j), n(i,j)) = \frac{\Gamma(i,j)}{n(i,j)E[r(i,j)] - \Gamma(i,j)} \qquad (10.1)$$

as the mean queue length at the link (i,j) and $[n(i,j)E[r(i,j)] - \Gamma(i,j)]^{-1}$ as the mean delay. Similarly, we can consider packet-loss probability if the buffer lengths are small. Using Lagrange multipliers one can accommodate constrained optimization (see Ref. 31 for more details).

The cost functions provided above may not be very good approximations of mean delay and queue lengths. Better approximations are provided in Ref. 31. However, it is an important direction for future research.

We consider the following joint routing and scheduling problem:
Find $n(i,j)$ and $\alpha_p(s,d)$ that minimizes

$$\sum_{(i,j)\in\mathcal{L}} f(\Gamma(i,j), n(i,j)) \qquad (10.2)$$

subject to

$$\Gamma(i,j) = \sum_{(s,d)} \sum_{p:(i,j) \text{ is on } p} \alpha_p(s,d)\lambda(s,d) \leq n(i,j)E[r(i,j)] \qquad (10.3)$$

$$0 \leq \alpha_p(s,d) \quad \text{for each } p, (s,d) \qquad (10.4)$$

and

$$\sum_p \alpha_p(s,d) = 1 \quad \text{for each } (s,d) \qquad (10.5)$$

where $\alpha_p(s,d)$ is the fraction of (s,d) traffic on route p, $\mathcal{L}$ is the set of links, and the inner summation in Equation 10.3 is overall possible routes for (s,d). The condition 10.3 is required to satisfy the stability condition at each link (i,j).

Obtaining the optimal solution in Equations 10.2 through 10.5 can be very time consuming because of the nonlinear cost function. Also, if it is not possible to satisfy the $\lambda(s,d)$ requirements of each (s,d), the above optimization problem may not provide any solution. Thus in the following, we first develop algorithms that will check for feasibility of the demands $\lambda(s,d)$. If these are not feasible, then we provide a solution that may be fair to all (s,d) pairs. Finally, we obtain a solution that is fair to all (s,d) pairs and optimizes the nonlinear cost function.

Consider the following optimization problem:

$$\text{max } \lambda \text{ such that} \qquad (10.6)$$

$$\sum_p \alpha_p(s,d) \geq \lambda \quad \text{for all } (s,d) \qquad (10.7)$$

and Equations 10.3 and 10.4 are satisifed where the summation in Equation 10.7 is overall possible paths p in the network. A solution to the above optimization problem can be considered fair and efficient. This is because if there is a routing and scheduling algorithm that satisfies all the traffic requirements $\lambda(s,d)$, then λ will be ≥ 1. If not, it provides the largest fraction of traffic that can be satisfied for each (s,d). This concept of fairness has also been considered in Refs. 25, 34 and 36. Furthermore, unlike Equations 10.2 through 10.5, this problem is a linear program (LP) and hence can be solved much faster than the nonlinear problem (Equations 10.2 through 10.5).

In addition to Equations 10.3 through 10.7, the network should also satisfy some transmission constraints. These constraints occur due to wireless nature of the links. In the 802.16 standard, these are given by stating that two links can be scheduled for transmission in the same slot if they are 1, 3, or 7 hops away from each other. However, in a practical system it may or may not be possible to schedule two links in a slot depending on the power of transmission, the distance between the receiving nodes, and other geographical factors. This can be decided in a particular scenario by actually taking measurements and finding the signal to interference and noise ratio (SINR) at different nodes. In the following, we will assume that this has been done for the network under consideration. Sometimes we can write these constraints as necessary or sufficient inequality constraints. For example, if no spatial channel reuse is allowed (as was done in the 802.16 standard in 2004 or if in the current standard we choose the option that spatial reuse is allowed only with a hop distance of seven in which case it may effectively be no spatial reuse) then necessary and sufficient conditions are

$$\sum_{(i,j)} n(i,j) \leq 1 \tag{10.8}$$

It is shown in Ref. 31 that if our transmission constraints are such that a node can receive successfully if and only if one of its neighboring nodes transmits in a slot and that a node can transmit only on one of its links at a time, then the necessary and sufficient conditions are

$$\sum_{j:(i,j)\in\mathcal{L}} n(i,j) \leq 1 \quad \text{for all } i$$

and

$$\sum_{i=(i,j)\in\mathcal{L}} n(i,j) \leq 1 \quad \text{for all } j \tag{10.9}$$

If we put the constraint that only one incoming or outgoing link at a node can be active at a time, then it is shown in Ref. 8 that necessary and sufficient conditions, in the context of WiMAX mesh networks, are

$$\sum_{j:(i,j)\in\mathcal{L}} n(i,j) + \sum_{j:(i,j)\in\mathcal{L}} n(i,j) \leq 1 \quad \text{for all nodes } i \tag{10.10}$$

It is argued in Ref. 8 that by using directional antennas and careful placement of nodes, these constraints can be quite realistic in the WiMAX.

Our general setup can work with transmission constraints of the type mentioned in Equations 10.9 and 10.10 along with the optimization problem considered above. The problem of transmission constraints has also been studied in Refs. 5, 17 and 25. Sometimes corresponding to these constraints, one may only be able to get sufficient inequality constraints. In the following, we will assume the transmission contraints have been put as linear inequality constraints and call them (T).

As mentioned above, a solution $n(i,j)$ and $\alpha_p(s,d)$ satisfying Equations 10.3, 10.4, 10.6, 10.7, and (T) will be considered efficient and fair. However, observe that the service provider will frequently need to run an algorithm in real time to obtain a solution and hence complexity of the algorithm will be an important issue. In general, the scheduling problem is *NP* hard because the

$n(i, j)$ needs to be integer valued (should be the number of slots in a frame assigned to link (i, j)). However, if we ignore the integrality of $n(i, j)$ and consider them as nonnegative fractions as considered above (Equations 10.3, 10.4, 10.6, 10.7), (T) becomes an LP problem that is computationally much more tractable. Once a solution $n(i, j), \alpha_p(s, d)$ is found then one can find an appropriate frame length T (in number of slots) such that $n(i, j)T$ is (approximately) integer valued. Then obtaining a schedule for the links is not difficult (see Ref. 31 for an algorithm).

If the number of nodes in the mesh is large, then complexity of the above LP can also be of concern because the number of variables $\alpha_p(s, d)$ can be exponential in number of nodes. However, in that case this LP can be reformulated in terms of link flows [3,5] and this problem can be taken care of. Another way to handle it, in terms of $\alpha_p(s, d)$ itself is by solving the dual problem as discussed in Ref. 24.

If λ obtained from the above optimization problem is ≥ 1, then there is a route and schedule for all (s, d) pairs and links that can satisfy the traffic requirements of all users. If $\lambda < 1$, then λ is the largest fraction of traffic requirements of all (s, d) pairs that can be satisfied by the network. Our solution of the above problem can also provide $\lambda > 1$. When this happens, the network has more bandwidth (BW)/throughput than needed to satisfy the current QoS requirements of all the users. Then the above solution allocates the extra resources to different (s, d) pairs in a fair way. The extra resources can be used by the TCP connections usefully because in our QoS requirements we have only specified the minimum mean throughput a TCP desires.

Next we find a solution that minimizes the cost function

$$\sum_{(i,j) \in \mathcal{L}} f(\Gamma(i, j), n(i, j)) \tag{10.11}$$

while satisfying Equations 10.3, 10.4, (T) and

$$\sum_p \alpha_p(s, d) \geq \lambda \quad \text{for all } (s, d) \tag{10.12}$$

where λ is the optimal solution obtained from LP, that is, Equations 10.3, 10.4, 10.6, 10.7, and (T). This is a nonlinear optimization problem and can be quite computationally intensive. Depending upon the actual form of f one can try to obtain efficient algorithms to compute an optimal solution. In Ref. 31, several algorithms have been identified that can be useful for functions of the form 10.1 (see, for example, Refs. 10 and 12). Furthermore, an LP can also be used if instead of minimizing *sum* of link costs Equation 10.11 we use minimization of $\max_{(i,j)} f(\Gamma(i, j), n(i, j))$ [31].

One can further improve the efficiency of the system if the optimal λ in Equation 10.6 is less than 1. In Ref. 31, a method is suggested where the fraction of demands satisfied for some of the (s, d) pairs can be increased without decreasing the fraction for other (s, d) pairs below the optimal λ obtained above.

The routing and scheduling provided above will ensure that the slot assignment $n(i, j)$ for link (i, j) is such that its average rate $n(i, j)E[r(i, j)]$ is sufficient to carry the overall traffic passing through it. However, it will not ensure that the throughput (rate) seen by traffic of a pair (s, d) will indeed get its required share of throughput. This may partly happen because the TCP connections passing through fewer hops tend to get more throughput than the other TCP connections sharing links with it (see Refs. 7 and 13). Thus to ensure that the traffic of some (s, d) pairs does not hog most of the throughput at a link, we will store the traffic of different (s, d) pairs in different queues at each link and provide the required throughput to each queue via WRR (weighted round robin). This will ensure that traffic of each (s, d) pair will get its share of throughput at each link on its routes.

In Sections 10.4 through 10.6, we show how the aggregate allocation of BW to the traffic of different (s, d) pairs will be used so as to ensure end-to-end QoS to individual flows. Section 10.4 considers the real-time traffic. Section 10.5 provides details for the TCP connections. Section 10.6 shows how to combine the real-time and data traffic to provide QoS to individual connections. In this section, we also verify via simulations that our QoS architecture actually provides the QoS. Our QoS architecture will be scalable in the sense that it will not require per flow processing at intermediate nodes.

We will observe in Section 10.6 that to accomodate the QoS architecture of Sections 10.4 through 10.6, some optimality of the solution provided above will be lost.

10.4   QoS for Real-Time Traffic

In this section, we design scheduling algorithms to guarantee QoS to individual UDP connections. Two important real-time applications are IP telephony and video conferencing. For these applications, the end-to-end delay of a packet should not exceed (say) 150 ms. If a packet exceeds this delay, it will be dropped. For satisfactory performance, the fraction of packets dropped for an application should be less than (say) 2%. To satisfy these QoS requirements, we propose that at the end of a (scheduling) frame we drop the packets that could not be transmitted through the wireless network. This will ensure a maximum delay of about 40 ms (for four frames of 10 ms each) in the wireless network (the rest of the delay margin is left for the remaining part of the network that a packet may have to travel). We develop algorithms, which will ensure that a particular user will not experience drop probability greater than 2%.

Packets generated by audio encoders (in IP telephony) usually generate a CBR traffic. But a video encoder (say MPEG) one may use in video conferencing generates VBR traffic (although downloading a stored video may arrive as a CBR traffic). To satisfy the QoS requirements of these two types of applications efficiently we require different considerations. Therefore, we consider these two cases separately. We consider scheduling of CBR traffic in Section 10.4.1 and VBR traffic in Section 10.4.2.

On the basis of the routing and scheduling algorithm of Section 10.3, we know the fraction $\alpha_p(s,d)$ of total average traffic requirement $\lambda(s,d)$ of each pair (s,d) passing through a route p. Then, as we will detail in Section 10.6, based on the average throughput requirement of each UDP and TCP connection of (s,d), we will decide which of the CBR, VBR, and TCP connections of (s,d) will pass through which route. Knowing the route that each connection will take, we decide the QoS architecture in the following to provide the QoS to each connection.

10.4.1 Scheduling of CBR Traffic

Let X (a constant) be the total amount of traffic generated during a frame by different CBR connections of a particular (s,d) pair following a particular route denoted by links $p_1, p_2, \ldots, p_h$ (this will be known on the basis of the algorithm in Section 10.3). As mentioned above, to provide delay guarantees to these flows, we propose dropping of data that cannot be transmitted at the end of the scheduling frame. Now, the scheduler has to ensure that the amount of data dropped conforms to the QoS requirements of the flow. Let the upper bound required on the drop probability of the packets of these flows be ϵ. (For simplicity of notation we are taking this upper bound same for all CBR applications. If different flows have different requirements then ϵ is the minimum of these requirements. Of course one can handle the general case in the same way for each flow separately.)

The scheduling problem for this CBR–UDP traffic is to calculate the number of slots $n_j, j = 1, \ldots, h$ required at link p_j such that X units of data can be transmitted to the MBS per scheduling frame and the end-to-end drop probability is bounded by ϵ.

We decompose the drop probability ϵ into $\{\epsilon_j, j = 1, \ldots, h\}$ such that $\prod_{j=1}^{h}(1 - \epsilon_j) \geq (1 - \epsilon)$. At link p_j the number of slots allocated for these flows has to ensure that the drop probability is upper bounded by ϵ_j. We use n_j for the allocation of slots for link p_j, $r(j)$ for $r(p_j)$, and X_j for $\prod_{k=1}^{j}(1 - \epsilon_k) X$ to simplify the notation. Then, n_j has to satisfy

$$\lim_{n \to \infty} \frac{\sum_{k=1}^{n} (X_j - n_j r_k(j))^+}{n X_j} \leq \epsilon_j \tag{10.13}$$

This reduces to $E((X_j - n_j r(j))^+) \leq \epsilon_j X_j$. We can rewrite it as

$$\int_0^{X_j/n_j} (X_j - n_j r) f_j(r) \, dr \leq \epsilon_j X_j \tag{10.14}$$

where $f_j(\cdot)$ is the pdf of the link rate $r(j)$, which is assumed to be known. The quantity on the left in Equation 10.14 is a nonincreasing function of n_j and hence it is easy to compute the required n_j.

Since the drop probability ϵ is small, we do not loss much in optimality if in the above calculation we replace X_j with X. Also, instead of arbitrarily choosing the values $\{\epsilon_j, j = 1, \ldots, h\}$, we can consider the optimization problem

$$\min \left\{ \sum_{j=1}^{h} n_j \right\}$$

subject to

$$\prod_{j=1}^{h} (1 - \epsilon_j) \geq (1 - \epsilon)$$

and

$$\int_0^{X_j/n_j} (X_j - n_j r) f_j(r) \, dr \leq \epsilon_j X_j, \quad j = 1, \ldots, h$$

The above allocation of slots to satisfy the QoS was proposed in Ref. 34. However, it was observed in Ref. 34 that even though one can obtain the required number $n(j)$ of slots needed to satisfy the QoS as above, in practice, due to very small probabilities ϵ, the number of slots needed actually becomes $X/r_{\min}(j)$ where $r_{\min}(j)$ is the minimum rate supported by link p_j. By taking $r_{\min}(j)$ as the minimum rate supported in the standard, $n(j)$ becomes independent of statistics of $r(j)$. This may be much easier to do and will entail only a small loss of optimality. Similar comments will hold for VBR scheduling in Section 10.4.2.

10.4.2   Scheduling of VBR Traffic

Consider K VBR flows generated at an (s, d) that will follow the same route $p_1, p_2, \ldots, p_h$. Let $D_n(k)$ be the amount of data generated by flow k in frame n. We assume that the arrival process $\{D_n(k), n \geq 0\}$ for each $k = 1, \ldots, K$ is stationary and ergodic with known statistics. We also assume that the arrival processes from the various sources are mutually independent. As in the case of CBR traffic, we provide delay guarantees to the VBR flows by dropping the data not transmitted by the end of each frame. The problem is to calculate the number of slots required by this VBR traffic on each node on its route in order to bound the drop probability by ϵ. Again obtain $\epsilon_j, j = 1, \ldots, h$ as in Section 10.4.1.

   The amount of resources required utilizes the statistics of the arrival process. Since the data not transmitted at the end of a frame is dropped we need to consider only the marginal distribution of $D_n(k)$ to calculate the amount of resources required at the first node. Also, since the drop probability is typically small, we can assume that the statistics of the arrival process is not distorted after flowing through the first node. Hence, we can use the same analysis for each of the nodes along the route. Choose ϵ_j^b and ϵ_j^d such that $(1 - \epsilon_j^b)(1 - \epsilon_j^d) \geq (1 - \epsilon_j)$. Now, find C_j such that

$$P\left(\sum_{k=1}^{K} D_1(k) > C_j\right) \leq \epsilon_j^b \tag{10.15}$$

This C_j/K is called the *equivalent bandwidth* of the VBR source [38]. If we take ϵ_j^b to be same for all j, C_j becomes independent of j, which simplifies the further design. Therefore we do that in the following and denote it by C. The MBS can treat a VBR source as a CBR flow generating C/K units of data per frame and calculate the number of slots required to satisfy the drop probability requirement of ϵ^d as in Section 10.4.

In practice, the exact statistics of a VBR arrival process may not be available. The statistics that is generally available is the maximum, the minimum, and the average data rates. To satisfy the QoS requirements of the flows, we can calculate the value of C by using a source model that has all the known characteristics of the original source but has the worst case behavior (i.e., gets the largest equivalent BW). It is shown in Ref. 20 that for the case of K independent, homogeneous, stationary sources with arrivals in a slot taking values in a finite set (this class covers Markov-modulated sources modulated by finite state Markov chains) the worst case drop probability is obtained by replacing these sources by independent identically distributed (i.i.d.) ON–OFF sources having the same maximum, minimum, and average rates.

Simulation results in Ref. 34 show that the slot allocation provided in Sections 10.4.1 and 10.4.2 does indeed provide QoS to individual flows.

10.5 QoS for TCP Traffic

This section develops scheduling algorithms that can guarantee the QoS for TCP connections. Some TCP applications, for example, email do not require any QoS. However, web traffic and file transfer may require certain minimum response time. We try to satisfy these QoS requirements by providing adequate minimum mean throughput to individual connections. But, if there is insufficient bandwidth to satisfy the minimum mean throughput of all the TCP connections, then again there is the question of how should we do the allocation in a fair way. In this case, unlike the UDP connections where we recommend admission control, one other option is to give these connections less bandwidth than requested (see more comments on this in Section 10.9).

Initially, we will consider the case of persistent TCP connections. These are long-lived connections that need to send a large file. The QoS requirement for these connections is the maximum response time. Later on we will also consider TCP–ON–OFF connections (see Ref. 16 for details on this model) that model the web traffic using HTTP 1.1. In this model, a TCP connection transfers multiple files. Between transfer of two files, a TCP connection may not have a file to transfer (OFF period) for sometime. This is the dominant traffic type in the current Internet. An appropriate QoS requirement for these connections is the mean file download time.

For both of the above TCP types the QoS can be satisfied if we ensure a minimum mean throughput to each connection. In the following, we provide scheduling schemes to ensure this. First, we compute the average through-put these connections need to satisfy their QoS. These computations can be used also to compute the total average throughput requirement of all TCP connections for any particular (s, d) that we need in Section 10.3.

We consider TCP persistent connections first. Let N^P persistent TCP connections of an (s, d) be passing through a particular route. Let λ_j^P be the minimum throughput requirements (in packets/s) and s_j^P the packet lengths (in bits) of the jth persistent TCP connection. Thus, the total average throughput requirement of the persistent TCP connections is $\lambda^P = \sum_{j=1}^{N^P} \lambda_j^P s_j^P$ bits/s.

We now consider TCP–ON–OFF traffic. Let N^O TCP–ON–OFF flows of (s, d) be following the same route as the N^P persistent connections mentioned above. For simplicity, assume all of them to have the same mean download time requirement of T^{on} (this is the mean time that will be taken to download a file by such a connection) and the same mean number of packets D to be down-loaded during an ON period. Let the packet size of a connection be s^O. Let the mean time between two downloads be T^{off}. Then the throughput required by such a TCP flow to satisfy its QoS requirement is $\lambda^m = Ds^O/T^{\text{on}}$. Also the long-term average throughput required by this flow is $\lambda^a = Ds^O/(T^{\text{on}} + T^{\text{off}})$. The probabilty of a connection being ON is $T^{\text{on}}/T^{\text{on}} + T^{\text{off}}$. Making the *assumption* that the ON–OFF processes of different connections are independent, the mean number of connections ON at anytime is $N^O T^{\text{on}}/T^{\text{on}} + T^{\text{off}}$. Now since each connection requires a throughput of λ^m when ON, the total throughput requirement of ON–OFF traffic at node i is

$$\lambda^O = \left(\frac{N^O T^{\text{on}}}{T^{\text{on}} + T^{\text{off}}} \right) \lambda^m = N^O \lambda^a$$

The above approximation improves as the number of connections N^O increases.

The overall average throughput needed for the persistent and ON–OFF connections is $\lambda_T = \lambda^P + \lambda^O$ bits/s.

Let $L = N^P + N^O$ TCP connections of an (s, d) be passing through a particular route as decided by the algorithm in Section 10.3. Suppose they have been guaranteed a mean throughput of λ_T bits/s at each node on its route (say via WRR as mentioned before). We will see below that this is not sufficient to guarantee minimum throughput to individual TCP connections passing through the same set of links. This will require extra care. First we consider persistent connections.

Consider the system shown in Figure 10.1. Let the N^P TCP connections are passing through (say) four queues. TCP$_i$ has window size W_i (assume

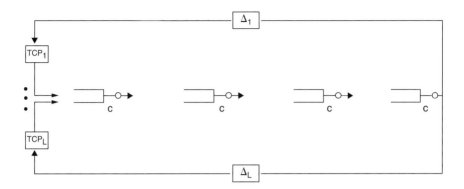

FIGURE 10.1
Multiple TCP flows through multiple queues with fixed rates.

it is fixed) and propagation delay Δ_i (representing delays in the rest of the network). At each queue the link speed is $c = \lambda^P$ bps (ensured say, by WRR discussed above). In this scenario, the packets/acks of different TCPs will be either at the first queue (and not at the other three queues) or propagating in the rest of the network (in propagation pipes Δ_is). If TCP$_i$ gets a throughput of λ_i packets/s, then by Little's law [37,38] it has on the average (under stationarity, proved in Ref. 16) $\lambda_i \Delta_i$ packets in the propagation pipe. Thus, the average queue length in the first queue is $\sum_{i=1}^{N^P} (W_i - \lambda_i \Delta_i)$ (actually it will be a little less than this; some packets might be getting serviced at the other queues). Since TCP$_i$ has packet length s_i^P (in bits) the mean queueing delay in the first queue is $\sum_{i=1}^{N^P} (W_i - \lambda_i \Delta_i) s_i^P / c$. As there is no queueing delays in the other queues, the total mean round trip time of TCP$_i$ is approximately $(1/c) \sum_{j=1}^{N^P} (W_j - \lambda_j \Delta_j) s_j^P + 3 s_i^P / c + \Delta_i$. Thus, the total throughput obtained by TCP$_i$ is

$$\frac{W_i c}{\sum_{j=1}^{N^P}(W_j - \lambda_j \Delta_j)s_j^P + 3s_i^P + \Delta_i c} \quad \text{(packets/s)} \tag{10.16}$$

To provide a desired throughput (or minimum throughput) to different TCPs, one needs to adjust c, W_j, s_j^P appropriately in Equation 10.16 such that it becomes equal to the desired throughput for each i (which may be different for different TCP connections). The bandwidth c (in bits/s) should equal $\sum_{i=1}^{N^P} \lambda_i^P s_i^P$ (if λ_i^P packets/s is the minimum mean throughput desired by connection i). It is possible that a service provider may not have the freedom to choose W_i and s_i^P for different TCP connections (these are selected by the receivers and the networks through which the connections are passing). However, even though the maximum window size W_i may not be controllable, if it is large enough, its mean size can be reduced to an appropriate size by dropping its packets in a controlled way (say) via random early detection (RED) [14] at the nodes in the mesh such that the required throughput can be provided to each TCP. We explain this in the following.

Let us fix a desired queueing delay of d^* seconds in the first queue. Define for each i, $\widetilde{\Delta}_i = \Delta_i + 3s_i^P/c$. We fix the desired mean window size of $E[W_i]$ such that

$$\frac{E[W_i]}{d^* + \widetilde{\Delta}_i} = \lambda_i^P \quad \text{for each } i \tag{10.17}$$

Now we use RED control for each TCP connection i and specify its RED parameters such that at average queue length d^*c, it will drop the packets of TCP_i with probability p_i, where

$$p_i = \frac{8}{(3(E_\pi[W_i] + 4)^2 + 5)} \quad \text{for each } i$$

(This has been used in Ref. 34 and is based on Ref. 27. It provides a reasonable approximation for small values of p_i. We are working on better approximations.) Then it can be shown (see Refs. 32 and 33) that this system will operate under steady state such that the first queue will have the mean queue length d^*c and each of the TCPs will have their mean window size $E[W_i]$ satisfying the above requirements. Furthermore, TCP_i will get the throughput λ_i^P packets/s. We will verify these claims via simulations in Section 10.7.

To include also the N^O TCP–ON–OFF connections in the above setup, we make $c = \lambda_T$ bps. Furthermore, we use RED to control the window size of a ON–OFF connection also where its mean window size $E[W]$ should satisfy Equation 10.17 with λ_i^P replaced by λ^m.

It is shown in Refs. 22 and 30 that the TCP connections can be grouped such that one needs only a few RED parameters to take care of the throughput requirements of different TCPs and per flow processing is not required.

Once we have ensured that the overall TCP traffic originating at the different nodes gets the bandwidth it requires at each node on its route, to ensure that the different persistent and ON–OFF TCP connections in it get the throughput they want, we set the window sizes according to Equation 10.17. To provide the needed mean window size, as explained above we can use RED at the bottleneck node along the route of the TCP connections. If the link rates are all fixed, then after ensuring that these flows get their required throughputs at each link, the node through which the TCP flows enter the mesh network is the bottleneck. However, due to the random variation of the link rates in wireless links, any link along the route can momentarily turn into a bottleneck (whenever its channel state is poor). To make our scheme work, one method is to implement RED control at all nodes along the path of the TCP flows. However, this involves difficulties in practical implementation since every node has to acquire the RED parameters of all the flows passing through it. An alternate method is to force the ingress node to be the bottleneck by providing about 3%–5% extra bandwidth to the flows at the other nodes along the route. This extra bandwidth whenever not used by these TCP flows can be provided to the best-effort traffic.

10.6   Joint Scheduling of UDP and TCP Flows

In this section, we address the problem of scheduling in presence of both UDP and TCP traffic. The requirements of UDP traffic are discussed in Section 10.4 and of TCP traffic in Section 10.5. From the arguments in these sections, in order to provide QoS to UDP we had to consider the worst case channel conditions whereas for TCP we had to consider the average channel rates. Thus, there is a huge difference between the average bandwidth requested and the average bandwidth provided to guarantee the QoS of the CBR and VBR connections. Here, we utilize this extra bandwidth for scheduling of TCP flows.

We provide priority to UDP traffic over TCP traffic in the network. It has been observed in Refs. 22, 30 and 34 that by doing this the delays experienced by UDP flows can be drastically reduced without affecting the throughput of the TCP flows. In the present context, it will allow us to save resources to provide QoS.

Let $\lambda_U(s,d)$ and $\lambda_T(s,d)$ be the average throughput required by the total UDP (for a CBR connection it is its rate, for a VBR connection, it is its equivalent bandwidth computed by taking h to be the maximum number of hops of a node from MBS) and TCP traffic generated by (s,d). Then we define $\lambda(s,d) = \lambda_U(s,d) + \lambda_T(s,d)$ as the average requirement of (s,d). We use this requirement in Section 10.3 to obtain the routing and scheduling for all the pairs. It is possible that the overall traffic of (s,d) is split over several routes. Let $\alpha_p(s,d)$ be the fraction of (s,d) traffic on route p.

If $\alpha_p(s,d) < 1$, then we will send $\alpha_p(s,d)$ fraction of $\lambda_U(s,d)$ and $\lambda_T(s,d)$ on route p. On a link (i,j) on p out of a total $n(i,j)$ slot assignment in Section 10.3, a fraction $n'(i,j,s,d)$ would be assigned for $\lambda(s,d)\alpha_p(s,d)$ traffic of (s,d), which is $\geq \lambda(s,d)\alpha_p(s,d)/E[r(i,j)]$.

This slot assignment is sufficient to satisfy the average throughput requirement of (s,d) traffic but may not be sufficient for the QoS of the real-time traffic. For this we do the following. First, from the above assigment of UDP and TCP traffic on route p we assign a number of UDP and TCP flows of (s,d) to the route p. Since we want to send the traffic of a particular flow on a single route (it is advisable not to split the traffic of a particular flow on more than one route), we may need to change a little bit the fraction $\alpha_p(s,d)$ of UDP and TCP traffic on route p so as to obtain an integral number of UDP and TCP flows of (s,d) on p. In the process, if an $\alpha_p(s,d)$ is very small, one may just make it zero. This can compromise a bit on the optimality of the solution but due to other benefits we would prefer such a solution.

Once the UDP and TCP flows of (s,d) to be routed on route p have been identified, we can compute the fraction of slots $n_U(i,j,s,d)$ needed on a link (i,j) on route p to satisfy the QoS of those CBR and VBR connections via the methods detailed in Section 10.4. Finally, we assign $n(i, j, s, d) \stackrel{\triangle}{=} \max(n_U(i, j, s, d), n'(i, j, s, d))$ fraction of slots of link (i, j) for the traffic of (s, d) passing through it. Also, we give priority to the real-time traffic of (s, d) over the TCP traffic of (s, d) on each link (i, j). Thus, since $n(i, j, s, d) \geq n_U(i, j, s, d)$, the QoS of the real-time traffic will be satisfied. Also, because $n(i, j, s, d) \geq n'(i, j, s, d)$, the long-term average rate of TCP and UDP traffic of (s, d) on (i, j) will be satisfied and hence the TCP traffic gets its share of throughput on each link (i, j). Often $n(i, j, s, d)$ will be much less than $n_U(i, j, s, d) + \lambda_T(s, d)\alpha_p(s, d)/E[r(i, j)]$ (the number of slots needed on link (i, j) to satisfy the QoS of TCP and UDP flows of (s, d) if we do not give priority to UDP flows) and hence this multiplexing of UDP and TCP traffic of (s, d) on (i, j) provides significant gains in resource requirement. At present the traffic mix in Internet is such that the real-time traffic makes less than 10% of the overall traffic. It is likely to be so in near future. In that case $n(i, j, s, d)$ will most likely be close to $n'(i, j, s, d)$ and hence the optimal solution obtained in Section 10.3 need not be modified.

In Figure 10.2, we show our overall QoS architecture when there are three nodes. Node i has N_{iU} UDP flows and N_{iT} TCP flows entering the mesh. Flows from nodes 1 and 3 enter node 2. At node 2 the flows from the three nodes are stored in separate queues and are provided their required throughput via WRR. In each queue the UDP packets are given priority over the TCP packets. The TCP flows are controlled via RED algorithms at their entry nodes.

We comment on the scalability of our overall scheme. At each node we need one queue for each source node whose traffic is passing through this node. In a mesh the number of nodes will not be large (it will be a few tens of nodes) and hence even in a large overall network this will not cause any problem. We also commented earlier that the RED control for TCP traffic can be used only at the node it enters. Also as in Refs. 22 and 30, we could classify TCP connections according to the mean window size (equivalently, their probability of loss) they need. Thus, even at the entry node we do not need to provide different

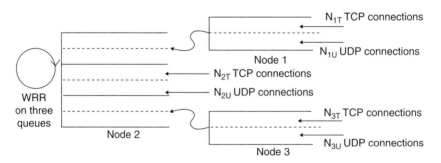

FIGURE 10.2
Overall scheduling of TCP and UDP flows to provide QoS. UDP flows get priority over TCP flows in each queue. RED applied on TCP flows at entry nodes. (We have separated UDP flows and TCP flows in each queue by a dotted line indicating each queue might be implemented via two queues.)

RED parameters for each TCP but rather for each TCP class. The number of TCP classes needed will not be more than 12 (as shown in Refs. 22 and 30 although one can work with even fewer). Furthermore, the computational complexity of our algorithms is rather low (at least if we ignore the nonlinear optimization in Section 10.3) and hence these can be implemented easily in real time for a reasonably large network.

10.7 Simulations

We consider the network shown in Figure 10.3. The network characteristics are summarized in Tables 10.1 and 10.2. The frame duration is 10 ms and scheduling is done over three frames. The channel rates of the channels vary randomly from one scheduling period (of three frames) to another independently. The link parameters shown in Figure 10.3 are the mean data rate per slot expressed in terms of the burst profile given in Table 10.2.

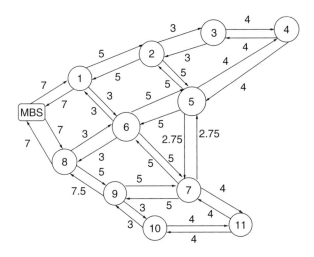

FIGURE 10.3
Network used in simulations.

TABLE 10.1

Physical Layer Parameters

Bandwidth	20 MHz
Number of subcarriers	256
Frame duration	10 ms
No. of OFDM symbols/frame	844
No. of OFDM symbols/minislot	4
Total no. of minislots/frame	211
No. of minislots/frame for uplink centralized scheduling	194

TABLE 10.2

Burst Profiles

Burst Profile No.	Modulation	Coding Rate	Uncoded Bytes per OFDM Symbol	Uncoded Bytes per Minislot
1	QPSK	1/2	24	6
2	QPSK	3/4	36	144
3	16QAM	1/2	48	192
4	16QAM	3/4	72	288
5	64QAM	2/3	96	384
6	64QAM	3/4	108	432

We consider the transmission constraints (Equation 10.9). For uplink, from each MSS node data need to be transmitted to the MBS. Each MSS node is sending traffic from three CBR sources with rates 16, 32, and 64 kbps. All MSSs (except MSS 5 and 1) are transmitting traffic from three VBR sources each transmitting with mean rate 256 kbps. The MSSs 5 and 6 have four VBR sources each, two transmitting at rate 256 kbps and the other two at rate 128 kbps. Each VBR source is a Markov-modulated source with transition matrix

$$\begin{pmatrix} 0.4 & 0.3 & 0.2 & 0.1 \\ 0.2 & 0.4 & 0.2 & 0.2 \\ 0.2 & 0.3 & 0.2 & 0.3 \\ 0.1 & 0.4 & 0.2 & 0.3 \end{pmatrix}$$

and with rates 73, 146, 293, and 586 kbps in the four states for a source with 256 kbps. For a source with 128 kbps in each state, the rate is half of that of a 256 kbps source. For uplink, for all MSSs (except MSSs 5 and 11) there are 24 persistent TCP connections with one requiring 16 kbps, seven requiring 64 kbps, eight requiring 128 kbps, and the rest 256 kbps. For the MSSs 5 and 11, there are 25 TCPs with two requiring 16 kbps, one requiring 32 kbps, six requiring 64 kbps, eight requiring 128, and the rest 256 kbps. For the down-link, each MSS is getting the same TCP traffic as for the uplink. There is no UDP traffic in the downlink. The packet sizes of the CBR connections are 100 bytes, of the VBR connections are 1500 bytes, and of the TCP connections are 1000 bytes. The TCP connections also have an extra propagation delay of 0.05 s.

From the above traffic details, the MSSs 1, 2, 3, 4, 8, 9, and 10 generate an average data of 4.5928 Mbps and the MSSs 5, 6, 7, and 11 generate, respectively, 4.5768, 4.592, 4.592, and 4.5768 Mbps in the uplink. Also the MSSs 1, 2, 3, 4, 8, 9, and 10 receive 3.7128 Mbps and MSSs 5, 6, 7, and 11 receive, respectively, 3.712, 3.6968, 3.6968, and 3.712 Mbps in the downlink. In computing these averages we have also included the acknowledgment traffic generated in the uplink and the downlink for the TCP traffic. The ack traffic in each queue is also given higher priority (along with the CBR and the VBR traffic) than the

TABLE 10.3

Performance of UDP Flows

	CBR Flows	VBR Flows
Max. avg. delay	28.01 ms	28.09 ms
Max. drop prob.	0.000%	0.000%

TABLE 10.4

Performance of TCP Flows

Percent Error	Number of Flows
<-25	0
-20 to -10	2
-10 to 0	85
0 to 10	351
10 to 20	77
20 to 30	9
>30	1

TCP data packets. For these average traffic requirements, we ran the LP, that is, Equations 10.3, 10.4, 10.6, 10.7, and 10.9 and obtained the $n(i,j)$ and $\alpha_p(s,d)$. Owing to lack of space we are not reporting these here. The λ obtained was equal to 1, that is, the network is able to satisfy the traffic demands of all (s,d) pairs. From $\alpha_p(s,d)$, we identified the CBR, VBR, and TCP connections of (s,d) that will use the route p. Also given the $n(i,j)$ for each link (i,j), using the algorithm in Ref. 31 we obtain the allocation of slots to different links in each scheduling period of three frames. Finally, we obtain via the WRR, the bandwidth allocation on each link for the traffic of each (s,d) passing through it and also the mean window size for each TCP connection.

The results obtained from these simulations are provided in Tables 10.3 and 10.4. Table 10.3 provides the maximum drop probability of 0.0 for the CBR and the VBR connections while the maximum average delay is 28.09 ms. Table 10.4 provides the error = Achieved average throughput − minimum required throughput for the different TCP connections. From these results, we see that the QoS of the real and data traffic is met very well. This has happened when most of the links are heavily loaded.

We have simulated several other networks with different traffic mixes and have made similar observations.

10.8 Admission Control

To provide QoS guarantees to different connections, admission control is required: if a new connection request arrives and the network does not have sufficient resources to provide it the QoS requested, the service provider should reject the request. A more relaxed rule would be: limit admission control decision (to reject) to applications with real-time hard constraints, for example, IP telephony and video conferencing. For other requests (e.g., audio/video streaming, web browsing) if there are insufficient resources, one can provide throughput less than requested by them. Of course here also a service provider may decide to reject an admission request if he is unable to provide beyond a certain fraction of requested throughput. In the following, we will limit ourselves to the first option: reject a request if its QoS cannot be satisfied.

Our admission control rule is simple and based on only the mean throughput requirements of the new TCP/UDP connection (for the VBR connection it is the equivalent BW). When a new connection request comes (say at (s, d)), it is sent to the MBS. The MBS chooses one of the routes p for which $\alpha_p(s, d) > 0$ at present (picking p with highest $\alpha_p(s, d)$ will be desirable). Then if λ bits/slot is the mean throughput requirement of the new request and (i, j) is a link on p, we compute $\bar{n}(i, j) = \lambda/E[r(i, j)]$, the number of slots required on link (i, j) to satisfy the mean throughput requirement. These requirements are summed up for all (i, j) on p and the MBS decides if it has the necessary number of slots (beyond the requirements of existing connections) to support the new request. If it has, the request is accepted. If not, then the MBS tries other routes with $\alpha_p(s, d) > 0$ till it gets one where the requirements are satisfied. If not, the request is denied. On first thought this procedure may look strange because one may expect that a route p on which $\alpha_p(s, d) = 0$ may also satisfy the QoS requirements. But this is unlikely because our routing and scheduling algorithm picks good routes to send the traffic of each (s, d). Of course the traffic of a new UDP connection will be given priority over the TCP connections of (s, d) on the route selected.

It is possible that if a routing and scheduling decision is being run for sometime, due to new connection arrivals or departures and other reasons, the current routing/scheduling may not be efficient. Thus, the MBS will occasionally run the routing and scheduling algorithm of Section 10.3 to reroute/reschedule the total traffic of different (s, d) pairs. It may periodically do it after every (say) M frames or when one of the following events happens:

a. A node/link fails.

b. The $E[r(i, j)]$ of some links (i, j) or the mean traffic requirements $\lambda(s, d)$ of some (s, d) pairs change drastically.

c. The admission control policy has been blocking too many calls in recent past.

10.9 Conclusions

In this chapter, we have designed efficient, fair, and practically implementable algorithms for routing and centralized scheduling in IEEE 802.16

mesh networks. We provide end-to-end QoS to different flows in the network. For this, we first provide an optimal and fair joint routing and scheduling solution to satisfy the aggregate mean traffic requirements of different source–destination pairs. Then we do scheduling at individual links to provide QoS to each flow. For this, we have handled UDP and TCP traffic separately at first and then jointly. Our algorithms are able to provide QoS to real- and nonreal-time individual flows efficiently and fairly. We have also provided an admission control policy, which is an important part of any QoS framework.

Acknowledgments

This work was partially supported by the DRDO-IISc program on Advanced Research in Mathematical Engineering. The simulations in this chapter have been done by Anil Kumar and Siddhartha Sankaran.

References

1. Air interface for fixed broadband wireless access systems, *IEEE STD 802.16*, October 2004.
2. Air interface for fixed and mobile broadband wireless access systems, *IEEE P802.16e/D12*, February 2005.
3. R. K. Ahuja, T. L. Magnanti, and J. B. Orlin, *Network Flows: Theory, Algorithms and Applications*, Englewood Cliffs, NJ: Prentice Hall, 1993.
4. I. F. Akyildiz, X. Wang, and W. Wang, Wireless mesh networks: A survey, *Computer Networks*, vol. 47, no. 4, pp. 445–487, 2005.
5. M. Alichery, R. Bhatia, and L. Li, Joint channel assignment and routing for throughput optimization in multiradio wireless mesh networks, in *Proc. Conf. Mobicom*, 2005.
6. M. Andrews and L. Zhang, Routing and scheduling in multihop wireless networks with time-varying channels, in *Proc. of the 15th Annual ACM-SIAM Symposium on Discrete Algorithms*, January 2004.
7. C. Barakat, E. Altman, and W. Dabbous, On TCP performance in hetero-geneous networks: A survery, *IEEE Communication Magazine*, vol. 38, pp. 40–46, January 2000.
8. M. Cao, V. Raghunathan, and P. R. Kumar, A tractable algorithm for fair and efficient uplink scheduling of multi-hop WiMax mesh networks, *Preprint, To appear in Proceedings of WiMesh 2006: Second IEEE Workshop on Wireless Mesh Networks*, Reston, VA, September 2006.
9. M. Cao, W. Ma, Q. Zhang, X. Wang, and W. Zhu, Modelling and performance analysis of the distributed scheduler in IEEE 802.16 mesh mode, *ACM MobiHoc*, 2005.
10. D. Z. Chen, O. Daescu, Y. Dai, N. Katoh, X. Wu, and J. Xu, Efficient algorithms and implementations for optimizing the sum of linear fractional functions with applications, *Journal of Combinatorial Optimization*, vol. 9, pp. 69–90, 2005.

11. W-P. Chen, C-F. Su, R. Rabbat, and T. Hamada, Asia-Pacific Network Operations & Management Symposium (APN-OMS 2005), September 2005. Radio resource management under fixed mobile convergence architecture.

12. W. Dinkelbach, On nonlinear fractional programming, *Management Science*, vol. 13, pp. 492–498, 1967.

13. M. Ergen, S. Coleri, and P. Varaiya, QoS aware adaptive resource allocation techniques for fair scheduling in OFDMA based broadband wireless access, *IEEE Transactions on Broadcasting*, vol. 49, no. 4, December 2003.

14. S. Floyd and V. Jacobson, Random early detection gateways for congestion avoidance, *IEEE/ACM Transactions Networking*, vol. 1, pp. 397–413, 1993.

15. L. Georgiadis, M. J. Neely, and L. Tassiulas, Resource allocation and cross-layer control in wireless networks, *Foundations and Trends in Networking*, vol. 1, no. 1, pp. 1–144, Now Publishers, 2006.

16. A. Gupta and V. Sharma, A unified approach for analyzing persistent, non-persistent and ON–OFF TCP sessions in the Internet, *Performance Evaluation*, vol. 63, pp. 79–98, 2006.

17. R. Gupta, J. Musacchio, and J. Walrand, Sufficient rate conditions for QoS flows in adhoc networks, *Preprint*, 2004.

18. M. Hawa and D. W. Petr, Quality of service scheduling in cable and broadband wireless access systems, in *10th IEEE International Workshop on Quality of Service*, pp. 247–255, May 2002.

19. C. Hoymann, Analysis and performance evaluation of the OFDM-based metropolitan area network IEEE 802.16, *Computer Networks*, vol. 49, pp. 341–363, 2005.

20. I. Hsu and J. Walrand, Admission control for multiclass ATM traffic with over-flow constraints, *Computer Networks and ISDN Systems*, vol. 28, pp. 1739–1751, 1996.

21. H. Jiang, W. Zhuang, X. Shen, A. Abdrabou, and P. Wang, Differentiated services for wireless mesh backbone, *IEEE Communication Magazine*, vol. 44, no. 7, pp. 113–119, July 2006.

22. V. Kamble and V. Sharma, A simple approach to provide QoS and fairness in Internet, in *Proc. International Conf. on Signal Processing and Communications (SPCOM 2004)*, Bangalore, December 2004.

23. D. Kim and A. Ganz, Fair and efficient multihop scheduling algorithm for IEEE 802.16 BWA systems, in *IEEE Broadnets'05*, Boston, 2005.

24. M. Kodialam and T. Nandagopal, Characterizing achievable rates in wireless multi-hop networks: The joint routing and scheduling problem, *ACM Mobicom*, September 2003.

25. M. Kodialam and T. Nandagopal, Characterizing the capacity region in multi-radio, multichannel wireless mesh networks, in *Proc. ACM Mobicom*, 2005.

26. X. Lin, N. B. Shroff, and R. Srikanth, A tutorial on cross-layer optimization in wireless networks, *IEEE Journal on Selected Areas in Communication*, vol. 24, pp. 1–12, August 2006.

27. J. Pandhye, V. Firoiu, D. Towsley, and J. Kurose, Modelling TCP throughput: A simple model and its empirical validation, in *ACM Proc. SIGCOMM'98*, 1998.

28. V. Raghunathan and P. R. Kumar, A counter-example in congestion control of wireless networks, *ACM Conference MSWiM*, 2005.

29. S. Ramanathan and E. L. Llyod, Scheduling algorithms for multihop radio networks, *IEEE/ACM Transactions on Networking*, vol. 1, pp. 166–177, April 1993.

30. V. Reddy, V. Sharma, and M. B. Suma, Providing QoS to TCP and real time connections in the Internet, *Queueing Systems*, vol. 46, pp. 461–480, 2004.

31. S. R. Sandeep, Joint routing and Scheduling for Multihop Wireless Networks, M.E. Thesis, Electrical Communication Engineering Dept, Indian Institute of Science, Bangalore, June 2006, (Also presented in poster session of *Conference on Stochastic Networks*, Urbana, IL, June 2006).

32. V. Sharma and P. Punyaslok, Stability and analysis of TCP connections with RED control and exogenous traffic, *Queuing Systems*, vol. 48, pp. 193–235, 2004.

33. H. Shetiya, *Efficient Routing and Scheduling Algorithms for IEEE 802.16 Mesh Networks*, M.E. Thesis, Dept. of ECE, Indian Institute of Science, Bangalore, 2005.

34. H. Shetiya and V. Sharma, Algorithms for routing and centralized scheduling to provide QoS in IEEE 802.16 mesh networks, in *Proc. of 1st ACM workshop on Wireless Multimedia Networking and Performance Modelling (WMuNeP 2005) in 7th International Symposium on Modelling, Analysis and Simulation of Wireless and Mobile systems (ACM/IEEE MSWiM 2005)*, October 2005.

35. H. Shetiya and V. Sharma, Algorithms for routing and centralized scheduling in IEEE 802.16 mesh networks, in *Proc. IEEE Wireless Communications and Networking Conference (WCNC06)*, April 2006.

36. V. Singh and V. Sharma, Efficient and fair scheduling of uplink and downlink in IEEE 802.16 OFDMA Networks, in *IEEE Wireless Communications and Networking Conference (WCNC06)*, April 2006.

37. J. Walrand, *An Introduction to Queueing Networks*, Prentice Hall, Englewood Cliffs, NJ, 1988.

38. J. Walrand and P. Varaiya, *High Performance Communication Networks*, San Francisco: Morgan Kaufmann, 1996.

39. H-Y. Wei, S. Ganguly, R. Izmailov, and Z. J. Haas, Interference aware IEEE 802.16 WiMax mesh networks, in *61st IEEE Vehiclular Technology Conf. (VTC 2005)*, Stockholm, 2005.

40. I. C. Wong, Z. Shen, B. L. Evans, and J. G. Andrews, A low complexity algorithm for proportional resource allocation in OFDMA systems, *Signal Processing Systems (SIPS)*, 2004.

41. S. Xu and T. Saadawi, Does the IEEE 802.11 MAC protocol work well in multihop wireless adhoc networks? *IEEE Communication Magazine*, vol. 39, no. 6, pp. 130–137, June 2001.

42. F. Xue and P. R. Kumar, Scaling laws for adhoc wireless networks: An information theoretic approach, *Foundation and Trends in Networking*, vol. 1, no. 2, pp. 145–270, Now Publishers, 2006.

Index

A

accessibility issues, Wi-Fi systems in
   emerging nations, 161–163
access points (APs)
   backhaul architecture
      device location model, 178–179
      network coverage, 188–193
         multihop connectivity, GW
         failures, 188–191
         overload avoidance AP/GW link
         design for, 191–193
         node/AP/GW neighbor distribution,
         181–184
         node distribution, single AP, 184–186
         overload conditions on, 192–193
         probability distribution GW/AP
         distances, 186–188
         radio propagation model, 179–180
         traffic load distribution, 191–193
   in WiMAX systems, 176
adaptive modulation
   IEEE 802.16a standard, 22–23
   rural area broadband services, 110–111
   WiMAX coding, 42
admission control, mesh topology, 215–216
advanced encryption standard-counter with
      CBC-MAC (AES-CCM),
      WiMAX security, 43–44
affordability issues
   fixed wireless infrastructures, emerging
      nations, 167–169
   Wi-Fi systems in emerging nations, 162
aggregate traffice, mesh topology, routing
      and scheduling of, 199–204
aggregation, MAC service data units, 62–63
air interfaces
   IEEE.802.16a specification, 11–12
   WiMAX architecture, 40–41
anytime-anywhere access, limitations in rural
      areas for, 105
architecture, WiMAX systems
   basic properties, 38–41
   flexibility, 44
   rural area broadband services, 106–108
ATM traffic, IEEE 802.16 standard, 2–3

automatic repeat request (ARQ)
   voice over internet protocol, enabling of,
      70–71
   WiMAX MAC frame format, 62
   WiMAX system, 64–65
availability issues, Wi-Fi systems in emerging
      nations, 162

B

backhaul architecture
   component connectivity, 181–188
      mathematical principles, 181
      neighbors distribution, 181–184
      node distribution, 184–186
      probability distribution, GW/AP
         distances, 186–188
   device location model, 178–179
   evolution of, 177–178
   future research issues, 192–193
   gateways and access points, 176–177
   network coverage, 188–193
      multihop connectivity, GW failures,
         188–191
      overload avoidance AP/GW link
         design for, 191–193
   radio propagation model, 179–180
   rural area broadband services, 111–112
   WiMAX usage scenario for, 9–10, 30
bandwidth requirements, fixed wireless
      access systems, 87–88
banking network connectivity, WiMAX
      systems, 49–50
base station (BS)
   connection confirmation and MPDU
      transmission, 64
   home access technology, 81–82
   IEEE 802.16 networking, 5–6
   rural area broadband services, 108
best-connected wireless service technology,
      rural area broadband services,
      112
bicycle-powered ICT systems
   remote WiMAX systems, 125–127
   rural telecommunications, 124–125
bit error ratios, adaptive modulation, IEEE
      802.16a standard, 23–24